TEXT BOOK
OF
PROBABILITY AND THEORETICAL DISTRIBUTIONS

DPH MATHEMATICS SERIES

TEXT BOOK OF PROBABILITY AND THEORETICAL DISTRIBUTIONS

By

A.K. Sharma

DISCOVERY PUBLISHING HOUSE
NEW DELHI-110002

Reprinted - 2019

First Published - 2005

ISBN: 978-81-7141-938-8

© Author

Text book of Probability and Thoretical Distribution

Published by:

DISCOVERY PUBLISHING HOUSE PVT. LTD.
4383/4B, Ansari Road, Darya Ganj
New Delhi-110 002 (India)
Phone: +91-11-23279245, 23253475; 43596065
E-mail: discoverybooksindia@gmail.com
discoverypublishinghouse@gmail.com
web: www.discoverypublishinggroup.com

Printed at:
Infinity Imaging Systems
Delhi

Preface

This book is written for B.A./B.Sc., students of all Indian Universities. The subject matter of the book is presented in a very straight forward manner so that the students are able to prepare this paper on their own and score well in examination.

We have tried to our best to keep the book free from the misprint. The author shall be grateful to the reader who point out errors and omissions which inspite of all care might have been there.

The author wish to express their thanks to the publishers Discovery Publishing House, New Delhi for bringing out this book in the present nice form.

A.K. Sharma

Contents

1

Probability and Expected Value

INTRODUCTION

(a) The word 'probability' or "chance" is very commonly used in day-to-day conversation and generally people have a ague idea about its meaning. For example, we come across statements like "Probably it may rain tomorrow". "It is likely that Mr. X may not come for taking his class today". "The chances of teams A and B wining a certain match are equal". "Probably you are right" : "It is possible that I may not be able to join you at the tea party". All these terms + possible, probably, likely, etc., convey the same sense, *i.e.*, the event is not certain to take place or, in other words, there is uncertainty about happening of the event in question. In layman's terminology the word probability," thus, connotes that there is uncertainty about the happening of the event. However, in mathematics and statistics we try to present conditions under which we can make sensible numerical statements about uncertainty and apply certain methods of calculating numerical values of probabilities and expectations.

(b) The modern theory of probability was developed by eminnot Russian mathematicians like Chebychev (1821-94). A Markov (1856-1922) and A.N. Kologorov. Kologorov axiomised the theory of probability and his book Foundations of Probability published in 1933 introduced probability and set function and is regarded as a classic.

(c) Starting with games of chance, 'probability' today has become one of the tools of statistics a fact, statistics and probability are so fundamentally interrelated that it is difficult to discuss statistics without an understanding of the meaning of probability. A knowledge of probability theory makes it possible to interpret statistical result, since many statistical procedures involve conclusions based on samples which are always affected by random variation, and it is by means of probability theory that we can express numerically the inevitable uncertainness in the resulting conclusions.

(d) The theory probability has its origin in the games of chance related to gambling such as throwing a die, tossing a coin, drawing cards from a pack of cards, etc. De Movire (1667-1754) contributed a lot to the subject and his work published in 1718 is the famous book. *The Doctrine of Chances.*

$$p(A) = \frac{\text{Number of favourable cases}}{\text{Total number of equally likely cases}}$$

For calculating probability we have to find out two things :

1. Number of favourable cases.
2. Total number of equally likely cases.

Ex., if a dice is thrown, the probability of obtaining an even number is $\frac{3}{6}$ or $\frac{1}{2}$ since three of the six equally possible results are even numbers.

Symbolically, if an event A can happen in 'a' ways out of a total of 'n' equally likely and mutually exclusive ways then the probability of occurrence of the event (called its success) † is denoted by :

$$p = \Pr(A) = \frac{a}{n}$$

and the probability of non-occurrence of the event †† (called its failure is given by :

$$q = \text{pr (Not A) or } P(\overline{A}) = \frac{n-a}{n} \text{ or } \frac{b}{n}$$

$$= 1 - \frac{a}{n} \text{ or } 1 - p \text{ or } 1 - \Pr(A).$$

Since the sum of the successful and unsuccessful outcomes is equal to the total number of events, we have

$$a + b = n$$

Dividing by n,

$$\frac{a}{n} + \frac{b}{n} = 1$$

so that $p + q = 1.$

Probability, therefore, may be written as a ratio. The numerator of the fraction corresponding to this ratio represents the number of successful (or unsuccessful) outcomes, while the denominator represents the total number of possible outcomes.

(e) Today the concept of probability has assumed great importance and the mathematical theory of probability has become the basis for statistical applications in both social and decision-making research. In fact, probability has become a part of our everyday life. In personal and management decisions, we face uncertainty and use probability theory, whether or not we admit the use of something so sophisticated. To quote Levin we live in a world in which we are unable to forecast the future with complete certainty. Our need to cope with uncertainty leads us to the study and use of probability theory.

PROBABILITY DEFINITION

The probability of a given event is an expression of likelihood or chance of occurrence of an event. A probability is a number which ranges from 0 (zero) to 1 (one) — zero for an event which cannot occur and 1 for an event certain to occur. How the number is assigned would depend on the interpretation of the term 'probability'.

Example:

From a bag containing 10 black and 20 white balls, a ball is drawn at random. What is the probability that it is black?

Solution:

Total number of balls in the bag = 10 + 20 = 30

Number of black balls = 10

Probability of getting a black ball or

$$p(A) = \frac{\text{Number of favourable cases}}{\text{Total number of equally likely cases}} \text{ or } \frac{a}{n}$$

$$= \frac{10}{30} = \frac{1}{3}$$

Probability of not getting a black ball or

$$q = \frac{20}{30} = \frac{2}{3}$$

Thus, $p + q = \frac{1}{3} + \frac{2}{3} = 1.$

Definition

The classical or mathematical or a priori definiton of probability as follows :

If there are s exhaustive, mutually exclusive and equally likely outcomes of a random experiment and r of them are favourable the happening of an event A, then the probability of the happening of A is

$$p(A) = \frac{r}{s}$$

It sometimes expressed as odd in favour of A are r to s –s or the odd against A are s – r to r.

Shortcomings of the Classical Approach

The classical definition of probability given above suffers from certain limitations. First, the definition cannot be applied whenever it is not possible to make a simple enumeration of cases which can be considered equally likely. For example, how does it apply to probability of rain? What are the possible cases ? We might think that there are two possibilities 'rain' or 'no rain'. But at any given time it will not usually be agreed that they are equally likely. Similarly, if a person jumps from the top of Qutab Minar the probability of his survival will not be 50 percent since survival and death, *i.e.*, the two mutually exclusive and exhaustive outcomes, are not equally likely.

Relative Frequency Theory of Probability

This classical definition is difficult or impossible to apply as soon as we deviate from the fields of coins, dice, cards and other simple games of chance. Secondly, the classical approach may not explain actual results in certain cases. For example, if a coin is tossed 10 times we may get 6 heads and 4 tails. The probability of a head is thus 0.6 and that of a tail 0.4. However, if the experiment is carried out a large number of times we should expect approximately equal number of heads and tails. As n increases, *i.e.*, approaches ∞ (infinity), we find that the probability of getting a head or tail approaches 0.5. The probability of an event can thus be defined as the relative frequency with which it occurs in an indefinitely large number of trials. If an event occurs a times out of n, its relative frequency is $\frac{a}{n}$ when n becomes infinity is called the limit of the relative frequency.

Symbolically, $$P(A) = \underset{n \to \infty}{Lt} \frac{a}{n}$$

Theoretically, we can never obtain the probability off an event a given by the above limit. However, in practice we can only try to have a close estimate convenience, the estimate of P(A) can be written as if it were actually P(A) and the relative frequency definition of probability may be expressed as :

$$P(A) = \frac{a}{n}$$

In the relative frequency definition the fact that he probability is the value which is approached by $\frac{a}{n}$ when n becomes infinity, emphasises a very important point, *i.e.*, probability involves a long-term concept. This means that if we toss a coin only 10 times, we may not get exactly 5 heads and 5 tails. However, as the experiment is carried out larger and larger number of times say, coin is thrown 10,000 times, we can expect heads and tails very closer to 50 percent.

The two approaches, classical and empirical, though seemingly same, differ widely. In the former, P(A) and $\frac{a}{n}$ were practically equal when n was large whereas in the latter we say that P(A) is the limit $\frac{a}{n}$ as n tends to infinity. In the second approach, thus, the probability itself is the limit of the relative frequency as the number of observations increases indefinitely.

The probability obtained by following relative frequency definition is called a posteriori or empirical probability as distinguished from a priori probability obtained by following the classical approach.

A clear distinction between a priori and empirical probability is quite important for proper understanding of the concept of probability. First a priori probability is normally encountered in problems dealing with games of chance—for example, dice and card. Normally, we think a of a priori probability as being deductive in nature, *i.e.*, from cause to effect and based on theory instead of the evidence of experience or experimentation.

Subjective Approach to Probability

The subjective approach to assigning probabilities was introduced in the year 1926 by Frank Ramsey in his book. The Foundation of Mathematics and other Logical Essays. The concept was further developed by Bernard Koopman, Richard Good and Leonard Savage. The subjective probability is defined as the probability assigned to an event by an individual based on whatever evidence is available. Hence such probabilities are based the beliefs of the person making the probability statement.

The application of personalistic concept to statistical problems has occurred virtually entirely in the post-World War II period, particularly in connection with statistical decision theory. This concept emphasises the fact that since probability of an event is the degree of belief or degree of confidence placed in the occurrence of an event by a particular individual based on the evidence available to him, different individuals may different in their degrees of confidence even when offered the same evidence. This evidence may on sit of relative frequency of occurrence data and any other quantitative or non-quantitative information. Persons might arrive at different probability assignments because of differences in values, experience and attitudes, etc. If an individual believes that it is unlikely that an event will occur, he will assign a probability close to zero to its occurrence. On the other hand, if he believes that it is very likely that the event will occur, he will assign a probability close to one.

The personalistic approach is very broad and highly flexible. It permits probability assignment to events for which there may be no objective data, or for which there may be a combination of subjective and objective data. However, one has to be very careful and consistent in the assignment of these probabilities otherwise the decisions made may be misleading. Used with care the concept is extremely useful in the context of situations in business decision-making.

Axiomatic Approach to Probability

The whole field of probability theory for calculations are based. The whole field of probability theory for finite sample spaces is based upon the following three axioms :

(1) The probability of the entire sample space is 1, *i.e.*, P(S) = 1.

(2) The probability of an event ranges from zero to one. In the event cannot take place its probability shall be zero and if it is certain, *i.e.*, bound to occur. Its probability shall be one.

(3) If A and B are mutually exclusive (or disjoint) events then the probability of occurrence of either A or B denoted by $P(A \cup B)$ shall be given by :

$$P(A \cup B) = P(A) + P(B)$$

It may be pointed out that out of the four interpretations of the concept of probability, each has it own merits and one may use whichever approach is convenient and appropriate for the problem under consideration. Experts disagree about which approach is the proper one to use.

CALCULATION OF PROBABILITY

Before discussing the procedure for calculating probability it is necessary to define certain terms as given below :

(a) **Experiment and Event :** The term experiment refers to describe and act which can be repeated under some given conditions. **Random** experiments are those experiments whose results depend on chance such a tossing of a coin, throwing of dice. The results of a random experiment are called outcomes. If in an experiment all the possible outcomes are known in advance and none of the outcomes can be predicted with certainty, then such an experiment is called a random experiment and the outcomes as events or chance event. Events are generally denoted by capital letters A, B, C, etc.

An event whose occurrence is inevitable when a certain random experiment is performed is called a certain r sure event. An event which can never occur when a certain random experiment is performed i called an impossible event.

An event which may or may not occur while performing a certain random experiment is known as a random event. Occurrence of 2 is a random event is the above experiment of the tossing of a dice.

(b) **Mutually Exclusive Events:** Two events are said to be mutually exclusive or incompatible when both cannot happen simultaneously in a ingle trial or, in other words, the occurrence of any one of them precludes the occurrence of the other. For example, if a single coin is tossed either head can be up or tail can be up, both cannot be up at the same time. Similarly, a person may be either alive or dead at a point of time—he cannot be both alive as well as dead at the same time. To take another example, if we toss a dice and observe 3, we cannot expect to also in the same toss of dice. Symbolically, if A and B are mutually exclusive event, P (AB) = 0.

DISJOINT SETS

(c) **Independent and Dependent Events :** Two or more events are said to be independent when the outcome of one does not affect, and is not affected by the other. For example, if a coin is tossed twice, the result of the second throw would in no way be affected by the result of the first throw. Similarly, the results obtained by throwing a dice are independent of the results obtained by drawing an ace from a pack of cards. To consider two events that are not independent, let A stand for a firm's spending a large amount of money on advertisement and B for its showing an increases in sales. Of course, advertising does not guarantee higher sales, but the probability that the firm will show an increase in sales will be higher if A ha taken place.

Dependent events are those in which the occurrence or non-occurrence of one event in any one trial affects the probability of other events in other trials. For example, if a card is drawn from a pack of playing cards and is not replaced, this will alter the probability that the second card drawn is, say an ace. Similarly, the probability of drawing a queen from a pack of 52 cards is $\frac{4}{52}$ or $\frac{1}{13}$. But if the card drawn (queen) is not replaced in the pack, the probability of drawing again a queen is $\frac{3}{51}$ ($\because$ the pack now contains only 51 cards out of which there are 3 queens).

(d) **Equally Likely Events :** Events are said to be equally likely when one does not occur more often than the others. For example, if an unbiased coin or dice is thrown, each face may be expected to be observed approximately the same number of times in the long run. Similarly, the cards of a pack of playing cards are so closely alike that we expect each card to appear equally often when a large number of drawings are made with replacement. However, if the coin or the dice is biased we should not expect each face to appear exactly the same number of times.

(e) **Simple and Compound Events :** In case of simple events we consider the probability of the happening or not happening of single event. For example, we might be interested in finding out the probability of drawing a red ball from a bag containing 10 white and 6 red balls. On the other hand, in case of compound event we consider the joint occurrence of two or more events. For example, if a bag contains 10 white and 6 red balls and if two successive draws of 3 balls are made, we shall be finding out the probability of getting 3 white balls in the first draw and 3 black balls in the second draw—we are thus dealing with a compound event.

(f) **Exhausted Events :** Events are said to be exhaustive when their totality includes all the possible outcomes of a random experiment. For example, while tossing a dice, the possible outcomes are 1, 2, 3, 4, 5 and 6 and hence the exhaustive number of cases is 6. If two dice are thrown once, the possible outcomes are :

(1, 1)	(1, 2)	(1, 3)	(1, 4)	(1, 5)	(1, 6)
(2, 1)	(2, 2)	(2, 3)	(2, 4)	(2, 5)	(2, 6)
(3, 1)	(3, 2)	(3, 3)	(3, 4)	(3, 5)	(3, 6)
(4, 1)	(4, 2)	(4, 3)	(4, 4)	(4, 5)	(4, 6)
(5, 1)	(5, 2)	(5, 3)	(5, 4)	(5, 5)	(5, 6)
(6, 1)	(6, 2)	(6, 3)	(6, 4)	(6, 5)	(6, 6)

The sample space of the experiment *i.e.*, 36 ordered pairs (6^2). Similarly, for a throw of 3 dice exhaustive number of cases will be 216 (*i.e.*, 6^3) and for n dice they will be 6^n.

Similarly, black and red cards are examples of collectively exhaustive events in a draw from a pack of cards.

(g) **Complementary Events :** Let there be two events A and B. A is called the complementary event of B (and vice versa) if A and B are mutually exclusive and exhaustive. For example, when a dice is thrown, occurrence of an even number (2, 4, 6) and odd number (1, 3, 5) are complementary events.

Simultaneous occurrence of two events A and B is generally written as AB.

ADDITIONAL THEOREMS OF PROBABILITY

The addition theorem states that if two events A and B are mutually exclusive the probability of the occurrence of either A or B is the sum of the individual probability of A and B. Symbolically.

$$P\ (A \text{ or } B) = P\ (A) + P\ (B).$$

Proof of the Theorem. If an event A can happen in a_1 ways and B is a_2 ways, then the number of ways in which either event can happen is $a_1 + a_2$. If the total number of possibilities is n, then by definition the probability of either the first or the second event happening is

$$\frac{a_1 + b_2}{n} = \frac{a_1}{n} + \frac{a_2}{n}$$

But $$\frac{a_1}{n} = P\ (A)$$

and $$\frac{a_2}{n} = P\ (B)$$

Hence $P\ (A \text{ or } B) = P\ (A) + P\ (B)$.

The theorem can be extended to three or more mutually exclusive events. Thus

$$P\ (A \text{ or } B \text{ or } C) = P\ (A) + P\ (B) + P\ (C).$$

Multiplication Theorem

This theorem states that if two events A and B are independent, the probability that they both will occur is equal to the product of their individual probability. Symbolically, if A and B are independent, then

$$P\ (A \text{ and } B) = P\ (A) \times P\ (B)$$

The theorem can be extended to three or more independent events.

Thus $\quad P\ (A, B \text{ and } C) = P\ (A) \times P\ (B) \times P\ (C).$

Proof of the Theorem: If an event A can happen in n_1 ways of which a_1 are successful and the event B can happen in n_2 ways of which a_2 are successful, we can combine each successful event in the first with each successful event in the second case. Thus, the total number of successful happenings in both cases is $a_1 \times a_2$. Similarly, the total number of possible cases is $n_1 \times n_2$.

Then by definition the probability of the occurrence of both events is

$$\frac{a_1 \times a_2}{n_1 \times n_2} = \frac{a_1}{n_1} \times \frac{a_2}{n_2}$$

But $\quad \frac{a_1}{n_1} = P\ (A)$

and $\quad \frac{a_2}{n_2} = P\ (B).$

$\therefore \quad P\ (A \text{ and } B) = P\ (A) \times P\ (B).$

In a similar way the theorem can be extended to three or more events

BAYES' THEOREM

The Bayes' theorem named after the British mathematician Rev. Thomas Bayes (1702–61) and published in 1763 in a short paper has became one of the most famous memoirs in the history of science and one of the most controversial. His contribution consists primarily of a unique method for calculating conditional probabilities. The so-called "Bayesian" approach to this problem addresses itself to the question of determining the probability of some event, A, given that another event, B, has been (or will be) observed, *i.e.*, determining the value of P (A/B). The event A is usually though of as sample information so that Bayes' rule is concerned with determining the probability of an event given certain sample information. For example, a sample output of 2 defectives in 50 trials (event A) might be used to estimate the probability that a machine is not working correctly (event B) or you might use the results of your first examination in statistics (event A) as sample evidence in estimating the probability of getting a first class (event B).

Bayes' theorem is based on the formula for conditional probability explained earlier, Let :

A_1 and A_2 = The set of events which are mutually exclusive (the two events cannot occur together) and exhaustive (the combination of the two events is the entire experiment; and

B = A simple event which intersects each of the A events as shown in the diagram below :

Observe the above diagram. The part of B which is within A_1 represents the area "A_1 and B" and the part of B within A_2 represents the area "A_2 and B".

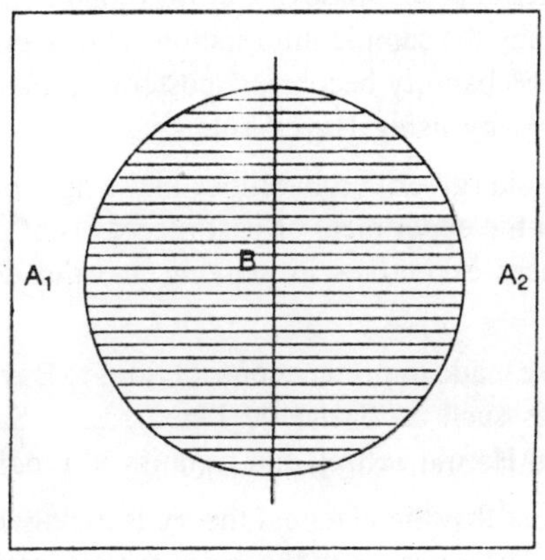

Then the probability of event A_1 given event B is

$$P(A_1/B) = \frac{P(A_1 \text{ and } B)}{P(B)}$$

and, similarly the probability of event A_2, given B, is

$$P(A_2/B) = \frac{P(A_1 \text{ and } B)}{P(B)}$$

where $P(B) = P(A_1 \text{ and } B) + P(A_2 \text{ and } B)$.

$P(A_1 \text{ and } B) = P(A_1) \times P(B/A_1)$, and

$P(A_2 \text{ and } B) = P(A_2) \times P(B/A_2)$

In general, let A_1, A_2, A_3, ..., A_i, ..., A_n be a set of n mutually exclusive and collectively exhaustive events. If B is another event such that P (B) is not zero, then

$$P(A_1 / B) = \frac{P(B / A_1) P(A_1)}{\sum_{i=1}^{k} P(B / A_1) P(A_1)}$$

Probabilities before revisions by Bayes' rule are called a priori or simply prior probability, because they are determined before the sample information is taken into account. A probability which has undergone revision in the light of sample information (via Bayes' rule) is called a posterior probability, since it represents a probability calculated after this information is taken into account.

Posterior probabilities are also called revised probabilities because they are obtained by revising the prior probabilities in the light of the additional information gained. Posterior probabilities are always conditional probabilities, the conditional event being the sample information. Thus a priori probability which is unconditional probability becomes a posterior probability, which is a conditional probability, by using Bayes' rule.

The revision of the old (given) probabilities in the light of the additional information supplied by the experiment or past records is of extreme help to business and management executives in arriving at valid decisions in the fact of uncertainty.

Several persons have made important contributions to Bayesian inference and decision procedures such as Bruno de Finetti, L. J. Savage. Howard Raiffa, Robert Schlaifer, Harold Jeffrey and Denniss V. Lindley.

It may be pointed out that the classical theory is mainly empirical since it employs only sample information as the basis for estimation and testing while the Bayeian approach employs any and all available information whether it is personal judgment or empirical evidence. Further, a Bayesian inference can be made on prior information alone or on both prior and sample information. The term 'prior information' implies the type of information which a statistician or a decision-maker has on an inferential problem before any sampling is conducted. Prior information often consists of personal judgements of the situation. Because of this the Bayesian method can actually be considered as an extension of the classical approach.

Some interesting points worth noting about Bayes' theorem are:

(1) The general conditional probability theorem asks. "What is the probability of the sample or experimental result given the state value"?

Whereas Bayes' theorem asks: "What is the probability of the event given the sample or experimental result"?

(2) When we talk of Bayes' theorem, different decision-makers may assign different probability to the same set of states of nature. Also we the preceding experiment as prior probabilities. As we proceed with repeated experiments, evidence accumulates and modifies the initial prior probabilities, thereby modifying the intensity of a decision-maker's belief in various states of nature. In other words the more evidence we accumulate, the less important are the prior probabilities.

(3) The notions of "priori" and "posterior" in Bayes' theorem are relative to a given sample outcome. That is, if a posterior distribution has been determined from a particular sample, this posterior distribution would be considered the prior distribution relative to a new sample.

The following example shall illustrate the application of Bayes' theorem:

SOLVED EXAMPLES

Example 1:

The Managing Committee of Vaishalli Welfare Association formed a sub-committee of 5 persons to look into electricity problem. Profiles of the 5 persons are :

1. male age 40

2. male age 43

3. female age 38

4. female age 27

5. male age 65

If a chairperson has to be selected from this, what is the probability that he would be either female or over 30 years ?

Solution:

P (female or over 30) = P (female) + P (over 30) – P (female and over 30)

$$= \frac{2}{5} + \frac{4}{5} - \frac{1}{5} = \frac{5}{5} = 1.$$

Example 2:

A problem in statistics is given to five students A, B, C, D and E. Their chances of solving it are $\frac{1}{2}, \frac{1}{3}, \frac{1}{4}, \frac{1}{5}$ and $\frac{1}{6}$. *What is the probability that the problem will be solved ?*

Solution:

Probability that A fails to solve the problem is $1 - \frac{1}{2} = \frac{1}{2}$

Probability that B fails to solve the problem is $1 - \frac{1}{3} = \frac{2}{3}$

Probability that C fails to solve the problem is $1 - \frac{1}{4} = \frac{3}{4}$

Probability that D fails to solve the problem is $1 - \frac{1}{5} = \frac{4}{5}$

Probability that E fails to solve the problem is $1 - \frac{1}{6} = \frac{5}{6}$

Since the events are independent the probability that all the five students fail to solve the problem is :

$$\frac{1}{2} \times \frac{2}{3} \times \frac{3}{4} \times \frac{4}{5} \times \frac{5}{6} = \frac{1}{6}$$

The problem will be solved if anyone of them is able to solve it.

$\therefore$ The probability that the problem will be solved $= 1 - \frac{1}{6} = \frac{5}{6}$

Example 3:

A man wants to marry a girl having qualities; white complexion—the probability of getting such a girl is one in twenty; handsome dowry—the probability of getting this is one in fifty; westernised manners and etiquettes—the probability here is one in hundred. Find out the probability of his getting married to such a girl when the possession of these three attributes is independent.

Solution:

Probability of a girl with white complexion

$$= \frac{1}{20} = 0.05$$

Probability of a girl with handsome dowry

$$= \frac{1}{50} = 0.02$$

Probability of a girl with westernised manners

$$= \frac{1}{100} = 0.01$$

Since the events are independent, the probability of simultaneous occurrence of all these qualities

$$= \frac{1}{20} \times \frac{1}{50} \times \frac{1}{100} = 0.05 \times 0.02 \times 0.01 = 0.0001.$$

If we are given n independent events A_1, A_2, A_3 ..., A_n with respective probability of occurrence as p_1, p_2, p_3, ..., p_n, then the probability of occurrence of at least one of the n events A_1, A_2, A_3, ..., A_n are can be determined as follows:

p (happening of at least one of the events) = 1

– p (happening of none of the events).

The following example hall illustrate the application of the above principle.

Example 4:

A person is known to hit the target in 3 out of 4 shots, whereas another person is known to hit the target in 2 out of 3 shots. Find the probability of the target being hit at all when they both try.

Solution:

The probability that the first person hits the target $= \frac{3}{4}$

The probability that the second person hits the target $= \frac{2}{3}$

The events are not mutually exclusive because both of them may hit the target.

P (AB) = P (A) P (B) since A and B are independent events.

The required probability $= \left(\frac{3}{4} + \frac{2}{3}\right) - \left(\frac{3}{4} \times \frac{2}{3}\right)$

$$= \frac{17}{12} - \frac{6}{12} = \frac{11}{12}.$$

Here we have applied the theorem

x P (A or B) = P (A) + P (B) – P (A and B).

Example 5:

Calculate the probability of picking a card that was a heart or a spade. Comment on your answer.

Solution:

Using the addition rule,

P (heat or spade) = P (heart) + P (spade)

$$- \text{P (heart and spade)} = \frac{13}{52} + \frac{13}{52} - \frac{0}{52} = \frac{26}{52} = \frac{1}{2}.$$

The probability that a card will be both a heart and a spade is zero since each individual card can be one and only one suit. The intersection in this case is non-existent called the null set because it contains no outcomes since heart and spade cannot occur simultaneously in the same card.

Example 6:

What is the probability of picking a card that was red or black?

Solution:

P (red or black) = P (red) + P (black)

Since there are 26 red and 26 black cards the required probability shall be

$$\frac{26}{52} + \frac{26}{52} = \frac{52}{52} = 1.0$$

The probability of red or black adds up to 1.0, this means that this is a certain event to happen.

Example 7:

One card is drawn from a standard pack of 52. What is the probability that it is either a king or a queen?

Solution:

There are 4 kings and 4 queens in a pack of 52 cards.

∴ The probability that the card drawn is a king $= \frac{4}{52}$

and the probability that the card drawn is a queen $= \frac{4}{52}$

Since the events are mutually exclusive, the probability that the card drawn is either a king or a queen

$$= \frac{4}{52} + \frac{4}{52} = \frac{8}{52} = \frac{2}{13}.$$

It is possible for both events to occur, the addition rule must be modified. For example, what is the probability of drawing either a king or a heart from a standard pack of cards? It is obvious that the events king and heart can occur together as we can draw a king of hearts (since king and heart are not mutually exclusive events). We must deduce from the probability of drawing either a king or a heart, the chance that we can draw both of them together. Hence for finding the probability of one or more of two events that are not mutually exclusive we use the modified from of the addition theorem.

P (A or B) = P (A) + P (B) – P (A and B)

P (A or B) = Probability of A or B happening when A or B are not mutually exclusive.

P (A) = Probability of A happening

P (B) = Probability of B happening

P (AB) = Probability of A and B happening together

In the example taken the probability of drawing a king or a heart shall be:

P (King or Heart) = P (King) + P (heart) + P (King and heart)

$$= \frac{4}{52} + \frac{13}{52} - \frac{1}{52} \text{ or } \frac{16}{52} = \frac{4}{13}.$$

In the case of three events.

P (A or B or C) = P (A) + P (B) + P (C) – P (AB)
– P (AC) – P (BC) + P (ABC).

Example 8:

A bag contains 30 balls numbered from 1 to 30. One ball is drawn at random. Find the probability that the number of the ball drawn will be a multiple of (a) 5 or 7, and (b) 3 or 7.

Solution:

The probability of the number being multiple of 5 is

$$P (5, 10, 5, 20, 25, 30) = \frac{6}{30}$$

The probability of the number being multiple of 7 is

$$P (7, 14, 21, 28) = \frac{4}{30}.$$

Since the events are mutually exclusive the probability of the number being a multiple of 5 or 7b will be

$$\frac{6}{30} + \frac{4}{30} = \frac{10}{30} = \frac{1}{3}$$

The probability of the number being multiple of 3 is

$$P (3, 6, 9, 12, 15, 18, 21, 24, 27, 30) = \frac{10}{30}$$

The probability of the number being multiple of 7 is

$$P\,(7, 14, 21, 28) = \frac{4}{30}$$

Since 21 is a multiple of 3 as well as 7, the drawing of the ball numbered 21 entails the occurrence of both the events and hence the probability of getting a number which is multiple of 3 or 7 is

$$= \frac{10}{30} + \frac{4}{30} - \frac{1}{30} = \frac{13}{30}.$$

CONDITIONAL PROBABILITY

The multiplication theorem explained above is not applicable in case of dependent events. Two events A and B are said to be dependent when B can occur only when A is known to have occurred (or vice versa). The probability attached to such an event is called the conditional probability and is denoted by P (A/B) or, in other word, probability of A given that B has occurred.

If two events A and B are dependent, then the conditional probability of B given A is :

$$P\,(B/A) = \frac{P(AB)}{P(A)}$$

Proof:

Suppose a_1 is the number of cases for the simultaneous happening of A and B out of $a_1 + a_2$ cases in which A can happen with or without happening of B.

$$\therefore \qquad P\,(B/A) = \frac{a_1}{a_1 + a_2} = \frac{a_1 / n}{(a_1 + a_2)/n} = \frac{P(AB)}{P(A)}$$

Similarly it can be shown that :

$$P\,(A/B) = \frac{P(AB)}{P(A)}.$$

The general rule of multiplication in its modified form in terms of conditional probability becomes :

$$P\,(A \text{ and } B) = P\,(B) \times P\,(A/B)$$

or

$$P\,(A \text{ and } B) = P\,(A) \times P\,(B/A)$$

For three events A, B and C, we have

$$P\,(ABC) = P\,(A) \times P(B/A) \times P\,(C/AB)$$

i.e., the probability of occurrence of A, B and C is equal to the probability of A times the probability of B given that A has occurred, time the probability of C given that both A and B have occurred.

Example 1:

A manufacturing firm produces units of a product in four plants. Define event A_i; a unit is produced in plant i, i =1, 2, 3, 4 and event B : a unit is defective. From the past records of the proportions f defectives produced at each plant the following conditional probabilities are set:

$$P(B/A_1) = 0.05,$$
$$P(B/A_2) = 0.10,$$
$$P(B/A_3) = 0.15,$$
$$P(B/A_4) = 0.02.$$

The first plant produces 30 percent of the units of the product, the second plant 25 percent, third plant 40 percent and the fourth plant 5 percent. A unit of the product made at one of these plants is tested and is found to be defective. What is the probability that the unit was produced in plant 3?

Solution:

We with to determine P (A_3/B). From the general form of the Bayes' theorem the probability is given by :

$$P(A_3 / B) = \frac{P(B / A_3)\ P(A_3)}{\sum_{i=1}^{k} P(B / A_i)\ P(A}$$

The computation of P (A_3/B) is shown below :

Plant Event	P (A)	P (B/A_i)	P (A_i) P (B/A_i)	$\frac{P(A_i)\ P(B / A_i)}{P(A_i) P(b / A_i)} \sum_{i=1}^{4}$
1	0.30	0.05	0.015	0.015/0.101 = 0.1485
2	0.25	0.10	0.025	0.025/0.101 = 0.2475
3	0.40	0.15	0.060	0.06/0.101 = 0.5941
4	0.05	0.02	.001	0.001/0.101 = 0.0099
	1.00		0.101 = P (B)	1.00

If is clear from the table that P (A_1 B) = 0.5941. The fourth column when summed gives that denominator of Bayes' theorem. From the table P (B) = 0.101, *i.e.*, the probability that a defective part is produced by this firm is 0.101. An advantage of preparing a table of the above type when employing Bayes' theorem is that the summing of probabilities in the last column to 1.0 (or close to 1 allowing for some numerical round-offs) gives us same assurance that the required probability has been correctly determined.

Example 2:

A company uses a 'selling aptitude test' in the selection of salesmen. Past experience has shown that only 70% of all persons applying for a sales position achieved a classification "dissatisfactory" in actual selling, whereas the remainder were classified as "satisfactory", 85% had scored a passing grade on the aptitude test. Only 25% of those classified unsatisfactory, had passed the test on the basis of this information. What is the probability that a candidate would be a satisfactory salesman given that he passed the aptitude test?

Solution:

If stands for a 'satisfactory' classification as a salesman and P stands for 'passing the test', then the probability that a candidate would be "satisfactory" salesman given that he passed the aptitude test is :

$$P(S/P) = \frac{(0.07)\ (0.85)}{(0.70)\ (0.85) + (0.30)\ (0.25)}$$

$$= \frac{0.595}{0.595 + 0.75} = 0.888$$

The result indicates that the tests are of value in screening candidates. Assuming no change in the type of candidates applying for the selling positions, the probability that a random applicant would be satisfactory is 70%. On the other hand, if the company only accepts an applicant if he passed the test, the probability increases to 0.888.

Example 3:

A bag contains 5 white and 3 black balls. Two balls are drawn at random one after the other without replacement. Find the probability that both balls drawn are black.

Solution:

Probability of drawing a black ball in the first attempt is

$$P\ (A) = \frac{3}{5+3} = \frac{3}{8}$$

Probability of drawing the second black ball given that the first ball drawn is black

$$P(B/A) = \frac{2}{5+2} = \frac{2}{7}$$

∴ The probability that both balls drawn are black is given by

$$P(AB) = P(A) \times P(B/A) = \frac{3}{8} \times \frac{2}{7} = \frac{3}{28}.$$

Example 4:

Find the probability of drawing a queen, a king and a knave in that order from a pack of cards in three consecutive draws, the cards drawn not being replaced.

Solution:

The probability of drawing a queen $= \frac{4}{52}$.

The probability of drawing a king after a queen has been drawn $= \frac{4}{51}$

The probability of drawing a knave given that a queen and king have been drawn $= \frac{4}{50}$

Since they are dependent events, the required probability of the compound event is :

$$\frac{4}{52} \times \frac{4}{51} \times \frac{4}{50} = \frac{64}{1,32,600} = 0.00048.$$

Example 5:

Assume that a factory has two machines. Past show that machine 1 produces 30% of the items of output and machine 2 produces 70% of the items. Further 5% of the items produced by machine 1 were defective and only 1% produced by machine 2 were defective. If a defective item is drawn at random, what is the probability that the defective item was produced by machine 1 or machine 2?

Solution:

Let A_1 = the event of drawing an item produced by machine 1.

A_2 = the event of drawing an item produced by machine 2,

and B = the event of drawing a defective item produced either by machine 1 or machine 2.

Then from the first information,

$P(A_1) = 30\% = 0.30$, $P(A_2) = 70\% = 0.70$

From the additional information

$P(B/A_1) = 5\% = 0.05$, $P(B/A_2) = 1\% = 0.1$

The required values are tabulated below :

Computation of Posterior Probabilities

(1) Events	(2) Prior probability P (A_i)	(3) Condition probability event B given event A P (BA_i)	(4) Join probability A (A_i and B) (2) × (3)	(5) Posterior (revised) probability P (A_i/B) (4) ÷ P (B)
A_1	0.30	0.05	0.015	0.015/0.022 = 0.682
A_2	0.70	0.01	0.007	0.007/0.002 = 0.318
Total	1.00		P (B) = 0.022	1.000

Without the additional information, we may be inclined to say that the defective item is drawn from machine 2 output since $P(A_2) = 70\%$ is larger than $P(A_2) = 70\%$ is larger tan $P(A_1) = 30\%$. With the additional information, we may give a better answer. The probability that the defective item was produced by machine 1 is 0.682 or 68.2% and that by machine 2 is only .318 or 31.8%. We may now say that the defective item is more likely drawn from the output produced by machine 1.

The above answer may be checked by actual number of items as follows:

If 10,000 items were produced by the two machines in a given period, the number of items produced by machine 1 is

$$10{,}000 \times 30\% = 3{,}000$$

and the number of items produced by machine 2 is

$$10{,}000 \times 70\% = 7{,}000.$$

The number of defective items produced by machine 1 is

$$3{,}000 \times 5\% = 150$$

and the number of defective items produced by machine 2 is

$$7{,}000 \times 1\% = 70$$

The probability that a defective item was produced by machine 1 is

$$= \frac{150}{150 + 70} = 0.682$$

and by machine 2 is $= \dfrac{70}{150 + 70} = 0.318.$

MATHEMATICAL EXPECTATION

The mathematical expectation (also called the expected value) of a random variable is the weighted arithmetic mean of the variable, the weights used to find the mathematical expectation are all the respective probabilities of the values that the variable can possibly assume.

If X denotes a discrete random variable which can assume the values $X_1, X_2, X_3, ...X_k$, with respective probabilities $p_1, p_2, p_3 ...., p_k$ where $p_1 + p_2, p_3 + ... + p_k = 1$ the mathematical expectation of X denoted by E(X) is defined as:

$$E(X)\ p_1 X_1 + p_2 X_2 + p_3 X_3 + ... + p_k X_k$$

Thus the expected value equals the sum of each particular value within the set (X) multiplied by the probability that X equals that particular value.

It should be noted that the concept of mathematical expectation was originally applied to games of chance and lotteries, but the notion of an expected value has become more generally applied and is now a common term in everyday parlance. Business situations frequently involve the consideration of expected values.

Example 1:

A petrol pump proprietor sells on an average Rs. 80,000 worth of petrol on rainy days and an average of Rs. 95,000 on clear days. Statistics from the Meteorological Department show that the probability is 0.76 for clear weather and 0.24 for rainy weather on coming Monday. Find the expected value of petrol sale on coming Monday.

Solution:

$$X_1 = 80{,}000\ p_1 = 0.24$$

$$X_2 = 80{,}000\ p_1 = 0.24$$

$$\Sigma(X) = p_1 X_1 + p_2 X_2 + \text{............}$$

$$= 0.24\ (80{,}000) + 0.76\ (95{,}000)$$

$$= 19{,}200 + 72{,}200 = \text{Rs. } 91{,}400$$

Thus the expected value of petrol sale on coming Monday is Rs. 91,400.

Example 2:

There are three alternative proposals before a businessman to start a new project:

Proposal A : Profit of Rs. 5 lakhs with a probability of 0.6 or a loss of Rs. 80,000 with a probability of 0.4.

Proposal B : Profit of Rs. 10 laksh with a probability of 0.4 or a loss of Rs. 2 lakhs a probability of 0.6.

Proposal C : Profit of Rs. 4.5 laksh with a probability of 0.8 or a loss of Rs. 50,000 with a probability of 0.2.

If he wants to maximise the profits and minimise the loss, which proposal should he prefer.

Solution:

We would calculate the mathematical expectation of each of the proposal.

$$E(X) = p_1 X_1 + p_2 X_2 + ... + p_k X_k$$

Proposal A : Expected Value = 5,00,000 (0.6) – 80,000 (0.4)

= 3,00,000 – 32,000 = 2,68,000.

Proposal B : Expected Value = 4,50,000 (0.8) – 50,000 (0.2)

= 3,60,000 – 10,000 = 3,50,000.

Since expected value is highest in case of proposal C, hence he should prefer proposal C.

Example 3:

A firm plans to bid Rs. 300 per tonne for a contract to supply 1,000 tonnes of a metal. It has two competitors A and B and it assumes that the probability that A will bid less than Rs. 300 per tonne is 0.3 and that B will bid less than Rs. 300 per tonne is 0.7. If the lowest bidder gets all the business and the firms bid independently, what is the expected value of the contract to the firm?

Solution:

There are two competitors A and B and the lowest bidder gets the contract. Value of Plan = 300 × 1,000 = Rs. 3,00,000

Contractor A : Probability that bid is less than Rs. 300 per tonne = 0.3

Probability that bid is Rs. 300 or more = 0.7

Contractor B : Probability that bid is less than Rs. 300 per tonne = 0.7

Probability that bid is Rs. 300 or more per tonne = 0.3

(1) If both bids less than Rs. 300

Probability is 0.3 × 0.7 = 0.21

Therefore plan value is 3,00,000 × 0.21 = 63,000

(2) If only A bids less than 300 and B bids more than 300,

Probability is $0.3 \times 0.3 = 0.09$

Therefore plan value is $3,00,000 \times 0.09 = 27,000$

(3) B bis less than 300 while A bids more than 300

Probability is $0.7 \times 0.7 = 0.49$

Therefore plan value is $3,00,000 \times 0.49 = 1,47,000$

Therefore value of plan is Rs. = 63,000 + 27,000 + 1,47,000

= 2,37,000

Hence expected value of plan is Rs. 2,37,000.

RANDOM VARIABLE AND PROBABILITY DISTRIBUTION

A random variable is also known as a chance variable or Stochastic variable. A random variable may be discrete or continuous. If the random variable takes on the integer values such as 0, 1, 2, ... then it is called a discrete random variable. If the random variable takes on all values within a certain interval then the random variable is called a continuous random variable. The amount of rainfall on a rainy day or in a rainy season the height and weight of individuals are examples of continuous random variable.

In terms of symbols if a variable X can assume discrete set of values $X_1, X_2, \ldots X_k$ with respective probabilities $p_1, p_2, \ldots p_k$ where $p_1 + p_2 + p_k = 1$, we say that a discrete probability distribution for X has been defined. The function P(X) which has the respective values $p_1, p_2 \ldots$ for $X = X_1, X_2 \ldots X_k$ is called the probability function or frequency function of X. The probability distribution of a pair of fair dice tossed is given below:

X	2	3	4	5	6	7	8	9	10	11	12
p(X)	1/36	2/36	3/36	4/36	5/36	6/36	5/36	4/36	3/36	2/36	1/36

where X denotes the sum of the points obtained. For example the probability of getting sum 4 is 3/36. Thus in 1,200 tosses of the dice we would expect 100 tosses to give the sum 4.

It should be noted that a probability distribution is analogous to relative frequency distribution with probabilities replacing relative frequencies. Thus we can think of probability distributions as theoretical or ideal limiting forms of relative frequency distribution when the number of observations is made very large. For this reason, we can think of probability distributions as being distributions for populations, whereas relative frequency distributions are distributions drawn from this population.

MISCELANEOUS EXAMPLES

Example 1:

A and B enter into a bet according to which A will get Rs. 200 if it rains on that day and will lose Rs. 100 if it does not rain. The probability of raining on that day is 0.7. What is the mathematical expectation of A?

Solution:

Expectation = $p_1x_1 + p_2x_2 + \ldots\ldots\ldots$

Probability of rain = 0.7

Probability of no rain = 1 – 0.7 = 0.3

Expectation = 200(.7) – 100(.3) = 140 – 30 = 110

Hence the mathematical expectation of A is Rs. 110.

Example 2:

The personnel department of a company has records which show the following analysis of its 200 engineers:

Age (Years)	***Bachelor's degree only***	***Master's degree***	***Total***
Under 30	*90*	*10*	*100*
30 to 40	*20*	*30*	*50*
Over 40	*40*	*10*	*50*
Total	*150*	*50*	*200*

If one engineer is selected at random from the company, find:

(a) the probability he has only a bachelor's degree,

(b) the probability he has a master's degree given that he is over 40;

(c) the probability he is under 30 given that he has only a bachelor's degree.

Solution:

Let us define the events A, B, C and D as follows:

A: an engineer has a bachelor's degree only

B: an engineer has a master degree

C: an engineer is under 30 years of age

D: an engineer is over 40 years of age.

(a) $P(A) = \frac{150}{200} = 0.75$

(b) $P(B/D) = \frac{P(B \cap D)}{P(D)} = \frac{10/200}{50/200} = 0.20$

(c) $P(C/A) = \frac{P(C \cap D)}{P(A)} = \frac{90/200}{150/200} = 0.60$

Example 3:

A certain production process produces items that are 10 percent defective. Each item is insected before supplying to customers but the inspector incorrectly classifies an item 10 percent of the time. Only items classified as good are supplied. If 820 items have been supplied in all, how many of them are expected to be defective?

Solution:

We have

$$P(D) = \text{Probability of a defective item} = \frac{10}{100} = 0.1$$

Also, $P(\text{classified as good/defective}) = \frac{10}{100} = 0.1$

$\therefore P(G) = \text{Probability of a good item} = 1 - 0.1 = 0.9$

and P(classified as good/good)

$= 1 - P(\text{classified as good/defective})$

$= 1 - 0.1 = 0.9$

$\therefore$ P(Defective/classified as good)

$$= \frac{P(D).\ P(\text{classified as good / defective})}{P(D).\ P(\text{classified as good / D}) + P(G).\ P(\text{classified as good / G})}$$

$$= \frac{0.1 \times 0.1}{0.1 \times 0.1 + 0.9 \times 0.9} = \frac{0.01}{0.82} = \frac{1}{82} = 0.012.$$

Example 4:

A salesman has a 60 percent chance of making a sale to each customer. The behaviour of successive customers is independent. If two customers A and B enter, what is the probability that the salesman will make a sale to A or B ?

Solution:

The probability that the salesman fails to make sale to A

$$= 1 - \frac{6}{10} = \frac{4}{10}$$

The salesman fails to make a sale to B

$$= 1 - \frac{6}{10} = \frac{4}{10}$$

Since the events are independent the probability that the salesman fails to make sales to both the customers.

$$= \frac{4}{10} \times \frac{4}{10} = \frac{16}{100}$$

$\therefore$ The probability that the salesman is able to make sales to A or B

$$= 1 - \frac{16}{100} = \frac{84}{800} = 0.84.$$

Example 5:

The records of 400 examinees are given below:

Score	*Educational Qualifications* **B.A.**	**B.Sc.**	**B.Com.**	***Total***
Below 50	*90*	*30*	*60*	*180*
Between 50 and 60	*20*	*70*	*70*	*160*
Above 60	*10*	*30*	*20*	*60*
Total	*120*	*130*	*150*	*400*

If an examinee is selected from this group of examinees, find

(i) the probability that he is a commerce graduate,

(ii) the probability that he is a science graduate, given that his score is above 60,

(iii) the probability that his score is below 50, given that he is a B.A.

Solution:

Let the various events be defined as follows:

A: selected examinee is a commerce graduate

B: selected examinee is a science graduate

C: selected examinee is a graduate with a score below 50.

(i) Required probability of commerce graduate, *i.e.*, $P(C) = \frac{150}{400} = \frac{3}{8}$

(ii) Required probability that he is a science graduate given that the score is above 60.

$$P(B/E) = \frac{P(B \cap E)}{P(E)} = \frac{30/400}{130/400} = \frac{3}{13}$$

(iii) Required probability that his below 50 and he is a graduate

$$P(D/A) = \frac{P(D \cap A)}{P(A)} = \frac{90/400}{120/400} = \frac{3}{4}$$

Example 6:

A manufacturing firm produces pipes in two plants l and ll with daily prodution 1,500 and 2,000 pipes respectively. The fraction of defective pipes produced by two plants l and ll are 0.006 and 0.008 respectively. If a pipe selected at random from that day's production is found to be defective, what is the chance that it has come from plant l, plant ll ?

Solution:

Let the probability of the possible events be:

$P(A_1)$ = Probability that a pipe is manufactured in plant l

$$= \frac{1{,}500}{1{,}500 + 2{,}000} = \frac{3}{7}$$

$P(A_2)$ = Probability that a pipe is manufactured in plant ll

$$= \frac{2{,}000}{1{,}500 + 2{,}000} = \frac{4}{7}$$

The given data and the necessary calculations can be summed up in the following table:

Event A_i	prior prob. $P(A_i)$	Conditional Prob. $P(E/A_i)$	Joint Prob. $P(Ai \cap E)$	Posterior (revised) Prob. $P(A_i/E)$
A_1	3/7	0.006	0.018/7	9/25 = 0.36
A_2	4/7	0.008	0.032/7	16/25 = 0.64

Since the probability is more in the second case, hence it is most probable that the defective pipe has been drawn from the output of the second plant.

Example 7:

The probability of X, Y and Z becoming managers are 4/9, 2/9 and 1/3 respectively. The probabilities that the Bonus scheme will be introduced if X, Y and Z become managers are 3/10, 1/2 and 4/5 respectively

(i) What is the probability that the bonus scheme will be introduced ?

(ii) If the Bonus scheme has been introduced, what is the probability that the manager appointed was X ?

Solution:

Let P(X) = Probability that X becomes manager

P(Y) = Probability that Y becomes manager

P(Z) = Probability that Z becomes manager

Let P(B/X) = Prob. that bonus scheme is introduced when X becomes manager. Similarly we can define P(B/Y) and P(B/Z)

We are given:

$P(X) = 4/9$

$P(Y) = 2/9$

$P(Z) = 1/3$

$P(B/X) = 3/10$

$P(B/Y) = 1/2$

$P(B/Z) = 4/5$

$P(B) = P(BX) + P(BY) + P(BZ)$

$$= \frac{4}{9} \times \frac{3}{10} + \frac{2}{9} \times \frac{1}{2} + \frac{1}{3} \times \frac{4}{5}$$

$$= \frac{12 + 10 + 24}{90} = \frac{46}{90} \text{ or } \frac{23}{45}$$

Using Bayes' theorem, the required probability is:

$$\text{S } P(X/B) = \frac{P(XB)}{P(B)} = \frac{P(X).\,P(B/X)}{P(B)}$$

$$= \frac{4/9 \cdot 3/10}{23/45} = \frac{60}{6 \times 23} = \frac{10}{23}$$

$$= \frac{12 \times 45}{23 \times 23} = \frac{6}{23}.$$

Example 7(a):

An industrial salesman wants to know the average of units he sells per sales call. He checks his past sales records and comes up with the following probabilities:

Sales in Units	*0*	*1*	*2*	*3*	*4*	*5*
Probability	*0.15*	*0.20*	*0.10*	*0.05*	*0.30*	*0.20*

What is the average number of units he sells per sales call ?

Solution:

The salesman wants to know the average number of units he sells per sales call. This is the same thing as saying that he wants to know the expected value of each sales call, where a sales call is the random variable x. The expected value is calculated by the formula :

$$E(x) = p_1x_1 + p_2x_2 + p_3x_3 + \dots\dots$$

$$= 0.15(0) + 0.20(1) + 0.10(2) + 0.05(3) + 0.30(4) + 0.20(5)$$

$$= 0 + 0.2 + 0.2 + 0.15 + 1.2 + 1.0 = 2.75$$

Thus he would expect to sell 2.75 or 3 units on each sales call.

Example 8:

A dealer in refrigerators estimates from his past experience the probabilities of his selling refrigerators sold in a day. These are as follows:

No. of refrigerators sold in a day	*0*	*1*	*2*	*3*	*4*	*5*	*6*
Probability	*0.03*	*0.20*	*0.23*	*0.25*	*0.12*	*0.10*	*0.07*

Find the mean number of refrigerators sold in a day.

Solution:

Mean number of refrigerators sold

$$= 0 \times 0.03 + 1 \times 0.2 + 0.23 + 3 \times 0.25 + 4 \times 0.12 + 5 \times 0.10 + 6 \times 0.07$$

$$= 0 + .2 + .46 + .75 + .48 + .5 + .42 = 2.81 .$$

Hence mean number of refrigerators sold in a day is 3.

Example 9:

Under an employment promotion programme it is proposed to allow sale of newspapers on the buses during off-peak hours. The vendor can purchase the newspapers at a special concessional rate of 25 paise per copy against the selling price of 40 paise. Any unsold copies are however, a dead loss. A vendor has estimated the following probability distribution for the number of copies demanded.

No. of copies	*15*	*16*	*17*	*18*	*19*	*20*
Probability	*.04*	*.19*	*.33*	*.26*	*.11*	*.07*

How many copies should be ordered so that his expected profit will be maximum ?

Solution:

Profit per copy = Selling Price – Purchasing Price.

= 40 Paise – 25 Paise = 15 Paise.

Expected Profit

Expected Profit = No. of copies × probability × Profit per copy.

Computation of Expected Profit

No. of copies	**Probability**	**Profit per copy**	**Expected Profit (paise)**
15	0.04	15	9
16	0.19	15	46
17	0.33	15	84
18	0.26	15	70
19	0.11	15	31
20	0.07	15	21

17 copies will give the maximum expected profit of 84 paise.

Example 10:

A company has two plants to manufacture scooters. Plant I manufactures 70% of the scooters and Plant II manufactures 30%. At plant I, 80% of scooters are rated standard quality and at plant II, 90 of scooters are rated standard quality. A scooter is picked up at random and is found to be of standard quality. What is the chance that it has come from plant I, or plant II ?

Solution:

Let A_1 = The event of drawing a scooter produced by plant I

A_2 = The event of drawing a scooter produced by plant II

E = Event of drawing a standard quality scooter produced either by plant I or plant II.

From the first information :

$P(A_1) = 70\% = 0.7,\ P(A_2) = 30\% = 0.3$

From the additional information :

$$P(E/A_1) = \frac{80}{100} = 0.80;\ P(E/A_2) = \frac{9}{100} = 0.90$$

$$P(E \cap A_1) = P(E/A_1) \times P(A_1) = 0.8 \times 0.7 = 0.56$$

$$P(E \cap A_2) = P(E/A_2) = P(E/A_2) \times P(A_2) = 0.9 \times 0.3 = 0.27$$

The probability that a standard quality scooter chosen at random is manufactured by plant I is given by the Bayes' theorem as :

$$P(A_1/E) = \frac{P(E/A_{1)}) \times P(A_1)}{P(E/A_1) \times P(A_1) + P(E/A_2) \times P(A_2)} = \frac{0.56}{0.56 + 0.27} = \frac{56}{83}$$

Similarly we can obtain $P(A_2/E)$

Example 11:

In an examination 30% students have failed in mathematics 20% of the students have failed in chemistry and 10% have failed in both mathematics and chemistry. A student is selected at random. What is the probability that

(i) *The student has failed in mathematics if it is known that he has failed in chemistry.*

(ii) *What is the probability that the student has failed in mathematics to chemistry ?*

Solution:

Let A and B denote the event that the student has failed in mathematics and chemistry respectively. Then we have:

$$P(A) = 0.30,\ P(B) = 0.20,\ P(AB) = 0.10$$

(i) Probability that the student selected at random has failed in mathematics if it is known that he has failed in chemistry.

$$= P(A/B) = \frac{P(AB)}{P(B)} = \frac{0.10}{0.20} = 0.50$$

(ii) Probability that the student selected at random has failed either in mathematics or chemistry = $P(A \cup B) = P(A) + P(B) - P(AB)$

$$= 0.30 + 0.20 - 1.0 = 0.40.$$

Example 12:

A committee of 4 people is to be appointed from 3 officers of the production department, 4 officers of the purchase department, two officers of the sales department and 1 chartered accountant. Find the probability of forming the committee in the following manner :

(a) There must be one from each category.

(b) It should have at least one from the purchase department.

(c) The chartered accountant must be on the committee.

Solution:

Total no. of officers = 3 + 4 + 2 + 1 = 10.

4 persons can be chosen out of 10 in $^{10}C_4$ ways

$$= \frac{10 \times 9 \times 8 \times 7}{4 \times 3 \times 2 \times 1} = 210 \text{ ways.}$$

(i) 1 Person may be chosen from each category in :

$$^3C_1 = 3,\ ^4C_1 = 4,\ ^2C_1 = 2,\ ^1C_1 = 1 \text{ ways}$$

∴ Probability of forming a committee having one from each category

$$= \frac{3 \times 4 \times 2 \times 1}{210} = \frac{24}{210} = 0.114$$

(ii) Let P(A) represent the probability of at least one purchase officer being chosen. Then P(B) is the probability of none being purchase officer, and in this case, all the four persons are to be taken from the other six persons not belonging to the purchase department, which may be done in

$$^6C_4 = {}^6C_2 = \frac{6 \times 5}{2 \times 1} = 15 \text{ ways}$$

$$\therefore P(B) = \frac{15}{210} = \frac{1}{14}$$

and $\quad P(A) = 1 - P(B) = 1 - \frac{1}{14} = \frac{13}{14}$

(iii) When the only chartered accountant is always included in the committee, we have to choose any three persons out of a total of 9 persons, which may be done in $^9C_2$ se

$$= \frac{9 \times 8 \times 7}{3 \times 2 \times 1} = 84 \text{ ways}$$

Hence the required probability $= \frac{84}{210} = \frac{2}{5}$

Example 13:

The daily production of a machine producing a very complicated item gives the following probabilities for the number of items produced:

$$P(A) = 0.20, \ P(2) = 0.35, \ P(3) = 0.45$$

Furthermore, the probability of defective items being produced is 0.02. Defectives are assumed to occur independently. Determine the probability of no defective during a day's production.

Solution:

Let A be the event that no defective item is produced during a day, then by total probability theorem

$$P(A) = P(1). \ P(A/1) + P(2). \ P(A/2) + P(3). \ P(A/3)$$

The probability that a defective item is produced is given to be 0.02, therefore, the probability that a non - defective item is produced will be 1– 0.02 = 0.98. Also defectives are assumed to occur independently, therefore

$$P(A/1) = 0.98, \ P(A/2) = (0.98)(0.98).$$

$$P(A/3) = (0.98)(0.98)(0.98)$$

$$P(A) = (0.20)(0.98) + (0.35)(0.98)^2 + (0.45)(0.98)^2$$

$$= 0.1960 + 0.3361 + 0.4322 = 0.9643.$$

Hence the probability of no defectives during a day's production is 0.9643.

Example 14:

A problem in statistics is given to two students A and B. The odds in favour of A solving the problem are 6 to 9 and against B solving the problem 12 to 10. If A and B attempt, find the probability of the problem being solved.

Solution:

$$P(\text{A's solving the problem}) = \frac{6}{6+9} = \frac{6}{15}$$

$$P(\text{B's solving the problem}) = 1 - \frac{12}{12+10} = \frac{5}{11}$$

$$P(\text{A's not solving the problem}) = 1 - \frac{6}{15} = \frac{9}{15}$$

P(Both A and B are unable to solve the problem)

$$= \frac{9}{15} \times = \frac{6}{11} = \frac{18}{55}$$

$$P(\text{the problem being solved}) = 1 - \frac{18}{55} = \frac{37}{55} = 0.673$$

Example 15:

An ordinary pack of cards is shuffled thoroughly and its colour is noted. The entire pack is shuffled thoroughly and again the top card is exposed and the colour noted. Five such successive observations have shown a red card, what is the probability that the next such exposure will give a black card - less than half, or more than half ? Why ?

Solution:

The probability that the exposure will give a black card shall be 1/2.

The reason for this answer is quite simple. There are 52 cards in a pack 26 red and 26 black. The probability of drawing a card of red (or black) colour randomly from the pack is 1/2. If we go on repeating the process with replacement, the probability remains the same, *i.e.*, 1/2 independent of the previous results. Hence, despite the fact that five successive observations have shown a red card, the probability of getting a black card in the sixth exposure will remain the same, *i.e.*, 1.2.

Example 16:

(i) A class consists of 80 students, 25 of them are girls and 55 boys, 10 of them are rich and remaining poor, 20 of them are fair complexioned. What is the probability of selecting a fair complexioned rich girl?

(ii) Explain why there must be a mistake in the following statement:

Solution:

(i) Probability of selecting

$$\text{a fair complexioned person} = \frac{20}{80} = \frac{1}{4}$$

$$\text{Probability of selecting a rich person} = \frac{10}{80} = \frac{1}{8}$$

The probability of selecting a girl $= \frac{25}{80} = \frac{5}{16}$

The probability of selecting a fair complexioned rich girl

$$= \frac{1}{4} \times \frac{1}{8} \times \frac{5}{16} = \frac{5}{512} = 0.0098$$

(ii) We know that probability of happening of an event can never exceed one. In the question given.

P(A) + P(B) + P(C)..........

$$= 0.11 + 0.23 + 0.37 + 0.16 + 0.09 + 0.05 = 1.01$$

Since the total probability is greater than one, there is some mistake in the statement given.

Example 17:

Three groups of workers contain 3 men and one woman, 2 men and 2 women, and 1 man and 3 women respectively. One worker is selected at random from each group. What is the probability that the group selected consists of 1 man and 2 women ?

Solution:

There are three possibilities in this case :

(i) Man is selected from the first group and women from 2nd and 3rd groups; or

(ii) Man is selected from the 2nd group and women from the 1st and 3rd groups; or

(iii) Man is selected from the 3rd group and women from 1st and 2nd groups.

∴ The probability of selecting a group of 1 man and 2 women

$$= \left(\frac{3}{4} \times \frac{2}{4} \times \frac{3}{4}\right) + \left(\frac{2}{4} \times \frac{1}{4} \times \frac{3}{4}\right) + \left(\frac{1}{4} \times \frac{1}{4} \times \frac{2}{4}\right)$$

$$= \frac{9}{32} + \frac{3}{32} + \frac{1}{32} = \frac{13}{32}$$

Example 18:

An artical manufactured by a company consists of two parts A and B. In the process of manufacture of part A, 9 out of 100 are likely to be defective. Similary, 5 out of 100 are likely to be defective in the manufacture of part B. Calculate the probability that the assembled part will not be defective.

Solution:

The assembled part will be defective if any of the part is defective

$\therefore$ The probability of the assembled part being defective.

$$= P[\text{that any of the part is defective}]$$

$$= P[A \cup B]$$

$$= P(A) + P(B) - P(AB)$$

$$= \frac{9}{100} + \frac{5}{100} - \frac{9}{100} \times \frac{5}{100}$$

$$= 0.09 + 0.05 - 0.0045 = 0.1355$$

$\therefore$ The probability that assembled part is not defective $= 1 - 0.1355$

$= 0.8645.$

Example 19:

A bag contains 6 white, 4 red and 10 blacks balls. Two balls are drawn at random. Find the probability that they will both be black.

Solution:

Total number of balls in the bag

$\therefore = 6 + 4 + 10 = 20$

Two balls can be drawn from 20 in ${}^{20}C_2$ ways

$$= \frac{20 \times 19}{2 \times 1} = 190 \text{ ways}$$

And 2 balls can be drawn from 10 blacks balls in ${}^{10}C_2$ or

$$= \frac{10 \times 9}{2 \times 1} 45 \text{ ways.}$$

$\therefore$ The probability that the two balls drawn at random are black

$$= \frac{45}{90} = 0.237$$

Example 20:

Three horses A,B and C are in a race. A is twice as likely to win as B and B is twice as likely to win as C. What are the respective probability of winning ?

Solution:

We have A : B : C = 4 : 2 : 1

Probability of winning for horse A = 4/7

Probability of winning for horse B = 2/7

Probability of winning for horse C = 1/7

Example 21:

An investment consultant predicts that the odds against the price of a certain stock will go up during the next week are 2 : 1 and odds in favour of the price remaining the same are 1 : 3. What is the probability that the price of the stock will go down during the next week?

Solution:

We have:

P (price of a certain stock not going up) = 2/3

P (price of a certain stock remaining same) = 1/4

∴ The probability that the price of the stock will go down during the next week

= p (price of the stock not going up and not remaining same)

= P (price of the stock not going up × P (price of the stock not remaining same)

$$= \frac{2}{3} \times \left(1 - \frac{1}{4}\right) = \frac{2}{3} \times \frac{3}{4} = \frac{1}{2} = 0.5$$

Example 22:

A bag contains 2 white and 3 black balls. Four persons A, B, C, and D in the order named each draw one ball and do not replace it The person to draw a white ball receives Rs. 200. Determine their expectations.

Solution:

Since only 3 black balls are contained in the bag, one person must win in the first attempt.

Probability that A wins = 2/5

$$\therefore \quad \text{A's expectations} = \frac{2}{5} \times 200 = \text{Rs. } 80.$$

Probability that A loses and B wins

$$= \frac{3}{5} \times \frac{2}{4} = \frac{3}{10}$$

$$\therefore \quad \text{B's expectation} = \frac{3}{10} \times 200 = \text{Rs. } 60.$$

Probability that A and B lose but C wins

$$= \left(\frac{3}{5}\right)\left(\frac{2}{4}\right)\left(\frac{2}{3}\right) = \frac{1}{5}$$

$$\text{C's expectation} = \frac{1}{5} \times 200 = \text{Rs.}40$$

Probability that A, B and C lose and D wins

$$= \left(\frac{3}{5}\right)\left(\frac{2}{4}\right)\left(\frac{1}{3}\right)\left(\frac{2}{2}\right) = \frac{1}{10}$$

$$\therefore \text{ D's expectation} = \frac{1}{10} \times 200 = \text{Rs.}20.$$

Thus the expectation of A, B, C and D respectively are Rs. 80, Rs. 60, Rs. 40 and Rs. 20.

Example 23:

A and B play for a prize of Rs. 1,000. A is to throw a dice first and is to win if he throws 6. If he fails B is to throw and is to win if he throws 6 or 5. If he fails, A is to throw again and to win if he throw 6, 5, or 4, and so on. Find their respective expectations.

Solution:

Probability of A's winning in the 1st throw

$= 1/6$

Probability of B's winning in the 2nd throw

$$= \frac{5}{6} \times \frac{2}{6} = \frac{5}{18}$$

Probability of A's winning in the 3rd throw

$$= \frac{5}{6} \times \frac{4}{6} \times \frac{3}{6} = \frac{5}{18}$$

Probability of B's winning in the 4th throw

$$= \frac{5}{6} \times \frac{4}{6} \times \frac{3}{6} \times \frac{4}{6} = \frac{5}{27}$$

Probability of A's winning in the 5th throw

$$= \frac{5}{6} \times \frac{4}{6} \times \frac{3}{6} \times \frac{2}{6} \times \frac{5}{6} = \frac{25}{324}$$

Probability of B's winning in the 6th throw

$$= \frac{5}{6} \times \frac{4}{6} \times \frac{3}{6} \times \frac{2}{6} \times \frac{1}{6} \times \frac{6}{6} = \frac{5}{324}$$

Total of A's chances of success

$$= \frac{1}{6} + \frac{5}{18} + \frac{25}{324} = \frac{169}{324}$$

Total of B's chances of success

$$= \frac{5}{18} + \frac{5}{27} + \frac{5}{324} = \frac{155}{324}$$

Their respective expectations are

$$A = \frac{169}{324} \times 1{,}000 = \text{Rs.}521.6$$

$$B = \frac{155}{324} \times 1{,}000 = \text{Rs.}478.4$$

Example 24:

A Bag contains 8 white and 4 red balls are drawn at random. What is the probability that 2 of them are red and 3 white ?

Solution:

Total number of balls in the bag = 8 + 4 = 12

Number of balls drawn = 5

5 balls can be drawn from 12 in $^{12}C_5$ ways, 2 red balls can be drawn from 4 red in $^4C_2$ ways and 3 white balls can be drawn from 8 white balls in $^8C_3$ ways.

∴ The number of favourable cases $^4C_2 \times {}^3C_3$ and the required probability is

$$P = \frac{^4C_2 \times {}^8C_3}{^{12}C_5}$$

$$= \frac{4 \times 3}{2 \times 1} \times \frac{8 \times 7 \times 6}{3 \times 2 \times 1} \times \frac{5 \times 4 \times 3 \times 2 \times 1}{12 \times 11 \times 10 \times 9 \times 8} = \frac{14}{33} = 0.424.$$

Example 25:

One bag contains 4 white and 2 black balls. Another contains 3 white and 5 black balls. If one ball is drawn from each bag, find the probability that (a) both are white, (b) are black, and (c) one is white and one is black.

Solution:

Probability of drawing a white ball from the first bag = 4/6

Probability of drawing a white ball from the second bag = 3/8

(a) Since the events are independent the probability that both the balls are white $= \frac{48}{6} \times \frac{3}{8} = \frac{1}{4}$

(b) Probability of drawing a black ball from the first bag = 2/6

Probability of drawing a black ball from the second bag = 5/8

Probability that both are black $= \frac{2}{6} \times \frac{5}{8} = \frac{5}{24}$

(c) That event "one is white one is black" is the same as event "either the first is white and the second is black or the first is black and the second is white."

$\therefore$ The Probability that one is white and one is black

$$= \left(\frac{4}{6}\right)\left(\frac{5}{8}\right) + \left(\frac{2}{6}\right)\left(\frac{3}{8}\right)$$

$$= \frac{20}{48} + \frac{6}{48} = \frac{13}{24}$$

Example 26:

A bag contains 5 white and 8 red balls. Two drawing of 3 balls are made such that

(a) the balls are replaced before the second trial, and

(b) the balls are not replaced before the second trial.

Find the probability that the first drawing will give 3 white and the second 3 red balls in each case.

Solution (a):

When balls are replaced:

Total number of balls in the bag = 5 + 8 = 13.

3 balls can be drawn from 13 in $^{13}C_3$ ways.

3 white balls can be drawn from 5 in $^{5}C_3$ ways.

3 red balls can be drawn from 8 in $^{8}C_3$ ways.

$\therefore$ The probability of drawing 3 white balls in the first trial is

$$= \frac{^5C_3}{^{13}C_3} = \frac{5}{143}$$

and the probability of 3 red balls at the second trial is

$$= \frac{^5C_2}{^{12}C_2} = \frac{28}{143}$$

∴ The probability of the compound event is

$$= \frac{5}{143} \times \frac{28}{143} = \frac{140}{20449} = 0.007$$

(b) When balls are not replaced:

At the first trial 3 white balls can be drawn in $^5C_3$ ways.

∴ The probability of drawing 3 balls at the first trial is

$$= \frac{^5C_3}{^{13}C_3} = \frac{5}{143}$$

When the white balls have been drawn and not replaced, the bag contains 2 white and 8 red balls. Therefore, at the second trial 3 balls can be drawn from 10 in $^{10}C_3$ ways and 3 red balls can be drawn from 8 in $^5C_3$ ways.

∴ The probability of 3 red balls in the second trial = $\frac{^8C_3}{^{10}C_3} = \frac{7}{15}$

∴ The probability of the compound event = $\frac{5}{143} \times \frac{7}{15} = \frac{7}{429} = 0.016.$

Example 27:

Five men in a group of 20 are graduates. If 3 men are picked out of 20 at random

(i) what is the probability that all are graduates, and

(ii) what is the probability of at least one being graduate ?

Solution:

Probability of finding one graduate is

$$= \frac{5}{20} = \frac{1}{4}$$

Probability of 1 graduate out of 3. selected

$$= {}^3C_1 \left(\frac{1}{4}\right)^1 \left(\frac{3}{4}\right)^2 = \frac{27}{64}$$

Probability of 2 graduates out of 3 selected

$$= {}^2C_2 \left(\frac{1}{4}\right)^2 \left(\frac{3}{4}\right)^1 = \frac{9}{128}$$

Probability of 3 graduates out of 3 selected

$$= {}^3C_3 \left(\frac{1}{4}\right)^3 \left(\frac{3}{4}\right)^0 = \frac{1}{64}$$

Therefore, the prob. of at least one graduate out of 3 selected

$$= \frac{27}{64} + \frac{9}{128} + \frac{1}{64} = \frac{65}{128}$$

Example 48:

A box contains 3 red and 7 white balls. One ball is drawn at random and in its place a ball of the other colour is put in the box. Now one ball is drawn at random from the box. Find the probability that it is red.

Solution:

At the second draw red ball can be drawn in two ways if at the first draw ball drawn is red or at the first draw the ball drawn is white.

When the first ball drawn is red:

The probability of drawing a red ball = 3/10

Now in the box a white ball is put in place of the red ball drawn so the box contains 2 red and 8 white balls.

Hence, the probability of drawing a red ball = 2/10

When the first ball drawn is white:

The probability of drawing a white ball = 7/10

Now in the box a red ball is put in place of the white ball drawn. There are thus 4 red and 6 white balls in the box.

$\therefore$ The probability of drawing a red ball = 4/10

Hence the probability of drawing a red ball

$$= \left(\frac{3}{10}\right) \times \left(\frac{2}{10}\right) + \left(\frac{7}{10}\right) \times \left(\frac{4}{10}\right)$$

$$= \frac{6}{100} + \frac{28}{100} = \frac{34}{100} = 0.34.$$

Example 29:

Prove that the sum of the probabilities of all possibilities in two independent events amount to certainty.

Solution:

Let the probability of success and failure in the first event be p_1 and q_1 and in the second event p_2 and q_2. Then,

The chance of success in the first event and success in the second event is

$p_1 \times p_2$

the chance of success in the first and failure in the second event is

$p_1 \times q_2$

the chance of failure in the first event and success in the second event is

$q_1 \times p_2$

the chance of failure in the first event and failure in the second event is

$q_1 \times q_2$

These are all the possibilities, and the sum of these possibilities is:

$= (p_1 \times p_2) + (p_1 \times q_2) + (q_1 \times p_2) + (q_1 \times q_2)$

$= p_1 (p_2 + q_2) + q_1 (p_2 + q_2)$

$= (p_1 + q_1)(p_2 + q_2)$

$= 1 \times 1 = 1$ ($\therefore p_1 + q_1 = p_2 + q_2 = 1$).

Example 30:

(a) *Given that four airlines provide service between Delhi and Mumbai. In how many distinct ways can a person select airlines for a trip from Delhi to Mumbai and back, if*

  (i) *he must travel both the ways by the same airline, and*

  (ii) *he travels from Delhi to Mumbai by one airline and returns by another?*

(b) *6 coins are tossed simultaneously. What is the probability that they will fall with 4 heads and 2 tails up ?*

(c) If A is the event of getting a prize on the first punch, and B the event of getting the prize on the second punch, calculate the probability of getting two prizes taking two punches on a punch board which contains 20 blanks and five prizes ?

(d) In how many ways can the word 'PROBABILITY' be arranged ?

(e) What is the probability of getting a number greater than two with an ordinary dice?

Solution (a):

(i) Since the person travels from Delhi to Mumbai and back by the same airline, therefore he can select any one airline out of four airlines. Hence the number of ways of selecting an airline is 4.

(ii) Since the persons travels from Delhi to Mumbai by one airline and returns by another so there are four ways of selecting one airline from Delhi to Mumbai. Now there are left three airlines by which he has to return. The number of ways of selecting an airline for return is 3.

Hence the number of ways of selecting airlines for a trip from Delhi to Mumbai such that both ways travels in different airlines.

$= 4 \times 3 = 12$

(b) The probability of a head = 1/2

The probability of a tail = 1/2

Out of 6 coins any 4 coins will show head and the remaining 2 will show tail.

$$\text{Hence the required probability} = {}^6C_4 \left(\frac{1}{2}\right)^4 \left(\frac{1}{2}\right)^2 = \frac{15}{64}$$

(c) Now first prize can be won in 5 ways. Then remain 4 prizes. The second prize can be won in 4 ways. Thus, there are $5 \times 4 = 20$ ways of winning two prizes

Since there are 25 punches on the punch board, there are 25 ways selecting first punch. Also there are 24 ways of selecting the second pouch.

Hence there are $25 \times 24 = 500$ ways of selecting two punches

∴ the probability of winning two prizes

$$= \frac{20}{600} = \frac{1}{30}$$

(d) There are 11 letters in which 'B' comes twice and also T comes twice if all the words were different then there were 11! ways of arranging them. Since 2 'B' and 'B'fs are coming they can be arranged in 2! × 2! ways.

Hence the total number of ways of arranging the word 'probability'

$$= \frac{11}{2! \times 2!} = 9979200$$

(e) In an ordinary dice 3, 4 and 6 means the number greater than 2.

Hence the probability of getting a number greater than 2 is 4/6 or 2/3.

Example 31:

What is the probability that a leap year, selected at random, will contain 53 Sunday ?

Solution:

A leap year consists of 366 days and, therefore, contains 52 complete weeks and 2 days extra. These 2 days may make the following 7 combinations:

1. Monday and Tuesday
2. Tuesday and Wednesday
3. Wednesday and Thursday
4. Thursday and Friday
5. Friday and Saturday
6. Saturday and Sunday
7. Sunday and Monday.

Of these seven equally likely cases only the last two are favourable. Hence the required probability = 2/7.

Example 32:

A husband and wife appear in an interview for two vacancies in the same post. The probability of husband's selection is 1/7 and that of wife's selection is 1/5. What is the probability that

(a) both of them will be selected

(b) only one of them will be selected, and

(c) none of them will be selected.

Solution:

Let A denote husband's selection and B denote wife's selection

(a) Probability that both of them will be selected. Since events are independent, we have

$$P(A \text{ and } B) = P(A)\,P(B) = (1/7) \times (1/5) = 1/35 = 0.029.$$

(b) Probability that only one of them will be selected

$$= P[(A \text{ and } \overline{B}) \text{ or } (B \text{ and } \overline{A})]$$

$$= P(A \text{ and } \overline{B}) + (B \text{ and } \overline{A})$$

$$= P(A)\,P(\overline{B}) = P(B)\,P(\overline{A})$$

$$= P(A)\,[1 - P(B)] + P(B)\,[1 - P(A)]$$

$$= \frac{1}{7}\left(1 - \frac{1}{5}\right) + \frac{1}{5}\left(1 - \frac{1}{7}\right)$$

$$= \left(\frac{1}{7} \times \frac{4}{5}\right) + \left(\frac{1}{5} \times \frac{6}{7}\right) = \frac{10}{35} = 0.286$$

(c) Probability that none of them will be selected

$$= P(\overline{A}) \times P(\overline{B}) = \frac{6}{7} \times \frac{4}{5} = \frac{24}{35}\ 0.686.$$

Example 33:

Out of 5 mathematicians and 7 statisticians, a committee consisting of 2 mathematicians and 3 statisticians is to be formed. In how many ways can this be done if

(a) any mathematician and any statistician can be included,

(b) one particular statistician must be on the committee,

(c) two particular mathematicians cannot be on the committee ?

Solution:

(a) 2 mathematicians out of 5 can be selected in ${}^5C_2$ ways.

3 statisticians out of 7 can be selected in ${}^7C_3$ ways.

Total number of possible selections = ${}^5C_2 \times {}^5C_3 = 10 \times 35 = 350$

(b) 2 mathematicians out of 5 can be selected in ${}^5C_2$ ways.

2 additional statisticians out of 6 can be selected in ${}^6C_2$ ways.

Total number of possible selections = ${}^5C_2 \times {}^6C_2 = 10 \times 15 = 150$

(c) 2 mathematicians out of 3 can be selected in ${}^3C_2$ ways.

3 statisticians out of 7 can be selected in $^7C_3$ ways.

Total number of possible selections = $^3C_2 \times {}^7C_3 = 3 \times 35 = 105$.

Example 34:

A market research firm is interested in surveying certain attitudes in a small community. These are 125 households broken down according to income, ownership of a telephone and ownership of a TV.

	Households with annual income of Rs. 8,000 or less		***Households with annual income above Rs. 8,000***	
	Telephone Subscriber	***No Telephone***	***Telephone Subscriber***	***No Telephone***
Own TV set	*27*	*20*	*18*	*10*
No TV set	*18*	*10*	*12*	*10*

(i) *What is the probability of obtaining a TV owner in drawing at random.*

(ii) *If a household has income over Rs. 8,000 and is a telephone subscriber, what is the probability that he has a TV ?*

(iii) *What is the conditional probability of drawing a household that owns a TV, given that the household is a telephone subscriber ?*

(iv) *Are the events 'ownership of a TV' and 'telephone subscriber' statistically independent ? Comment.*

Solution:

(i) Probability of drawing a TV owner at random, *i.e.*,

$$P = \frac{\text{Number of favourable cases}}{\text{Total number of equally likely cases}}$$

$$= \frac{75}{125} = 0.6$$

(ii) There are 30 persons (18 + 12) whose household income is above Rs. 8,000 and are also telephone subscribers. Out of these 18 own TV set. Hence the probability of this group of persons having a TV set.

$$P = \frac{18}{30} = 0.6$$

(iii) Out of 75 (27 + 18 + 18 + 12) households who are telephone subscribers, 45 (27 + 18) households have TV sets. Hence the conditional probability of drawing a household that owns a TV given that the household is a telephone subscriber.

$$P = \frac{45}{75} = 0.6$$

(iv) Two events A and B are statistically independent if

$P(AB) = P(A) \times P(B)$

Let A denote those who have TV and B those who are telephone subscribers. Probability of a person owning a TV

i.e., P(A). $P = \dfrac{75}{125}$ [$\because$ persons out of a total of 125 own a TV]

Probability of a person being a telephone subscriber

$P = \dfrac{75}{125}$ [$\because$ 75 persons out of a total of 125 are telephone subscriber]

$$P(AB)\ P = \frac{45}{125} = \frac{9}{25}$$

$$P(A) \times P(B)\ P = \frac{75}{125} \times \frac{75}{125} = \frac{9}{25}$$

Hence $P(AB)\ P = P(A) \times P(B)$

We, therefore, conclude that the events 'ownership of a TV' and 'telephone subscriber' are statistically independent.

Example 35:

A University has to select an examiner from a list of 50 persons. 20 of them women and 30 men, 10 of them knowing Hindi and 40 not, 15 of them being teachers and the remaining 35 not. What is the probability of the University selecting a Hindi-knowing woman teacher ?

Solution:

Probability of selecting a woman = 20/50

Probability of selecting a teacher = 15/50

Probability of selecting a Hindi-Knowing candidate = 10/50

Since the events are independent the probability of the University selecting a Hindi-knowing woman teacher

$$= \frac{20}{50} \times \frac{15}{50} \times \frac{10}{50} = \frac{3}{125} = 0.024$$

Example 36:

The probability that a boy will get a scholarship is 0.9 and that a girl will get is 0.8. What is the probability that at least one of them will get the scholarship ?

Solution:

The probability that a boy will get a scholarship = 0.9

The probability that a girl will get a scholarship = 0.8

The probability that at least one of them will get the scholarship is

$P(A) + P(B) - P(AB)$

$= .9 + .8 - (.9 \times .8)$

$= 1.7 - .72 = 0.98.$

Example 37 :

(a) A can solve 90 percent of the problems given in a book and B can solve 70 percent. What is the probability that at least one of them will solve a problem selected at random ?

(b) In a single throw of two dice, what is the probability of obtaining a total of at least 10 ?

Solution:

(a) Probability that A will not be able to solve the problem

$$= 1 - \frac{9}{10} = \frac{1}{10}$$

Probability that B will not be able to solve the problem

$$= 1 - \frac{7}{10} = \frac{3}{10}$$

Probability that none of them will be able to solve the problem

$$= \frac{1}{10} \times \frac{3}{10} = \frac{3}{100}$$

Hence the probability that at least one of them will solve the problem

$$= 1 - \frac{3}{100} = \frac{97}{100}$$

(b) Two dice can be thrown together in $6 \times 6 = 36$ ways.

We have to find the probability of getting a sum of at least ten, *i.e.*, either ten, eleven or twelve.

The probability of getting a total 10 in a single throw of two dice (6,4) (4,6) (5,5)

$= 3/36$

The probability of getting a total 11 in a single throw of two dice (6,5) (5,6)

$= 2/36$

Probability of getting a total 12 in a single throw of two dice (6,6)

$= 1/36$

Since the events are mutually exclusive the probability of obtaining a sum of at least 10 in a single throw of two dice $= \frac{3}{36} + \frac{2}{36} + \frac{1}{36} = \frac{6}{36} = \frac{1}{6}$

Example 37(a):

A Bag contains 10 white and 6 black balls. 4 balls are successively drawn out and not replaced. What is the probability that they are alternately of different colours ?

Solution:

(a) Beginning with white:

The probability of drawing a white ball = 10/16

The probability of drawing a black ball then = 6/15

The probability of drawing a white ball then = 9/14

The probability of drawing a black ball then = 5/13

Therefore, the probability of the compound event

$$= \frac{10}{16} \times \frac{6}{15} \times \frac{9}{14} \times \frac{5}{13} = \frac{45}{728}$$

(b) Similary, beginning with black:

The probability of drawing a black ball = 6/16

The probability of drawing a white ball then = 10/15

The probability of drawing a black ball then = 5/14

The probability of drawing a white ball then = 9/13

The probability of the compound event

$$= \frac{6}{16} \times \frac{10}{15} \times \frac{5}{14} \times \frac{9}{13} = \frac{45}{728}$$

The above two events are mutually exclusive. Therefore, the required probability that 4 exclusive drawn balls are alternately of different colours (without mentioning the colour with which to begin).

$$= \frac{45}{728} + \frac{45}{728} = \frac{90}{728} = 0.124.$$

Example 38:

In a product testing procedure, each radio on an assembly line must pass two inspection points before being packaged for shipment. The probability is p1 = 0.7 that a defective radio is detected at the first inspection point and p2 = 0.8 that a defective radio is detected at the second inspection point. What is the probability that a defective radio will be packaged for shipment ?

Solution:

A defective radio has to pass two inspection points before it is packaged. The probability that such a radio will pass the first inspection point is P(A) = (1 – P1) = 0.03. The probability of passing the second inspection point given that the defective radio passed the first point, is P(B/A) = (1 – P2) = 0.20. The probability that a defective radio will pass both inspection points is, therefore,

$$P(A \text{ and } B) = (1 - P1) = (1 - P2) = (0.30)\,(0.20) = .06$$

We conclude that about 6 percent of the defective radios will be packaged for shipment under the present inspection plan.

Example 39:

A manufacturing firm produces steel pipes in three plants with daily production volumes of 500, 1,000 and 2,000 units respectively. According to past experience, it is known that the fraction of defective output produced by the three plants are respectively .005, .008 and 0.10. If a pipe is selected from a day's total production and found to be defective, find out

(i) From which plant the pipe comes ?

(ii) What is the probability that it came from the first plant ?

Solution:

(i) According to Bayes' theorem we have from the problem the following events:

A1: Production volume of first plant = 500 units

A2: Production volume of second plant = 1,000 units

A3: Production volume of third plant = 2.000 units

E: a defective item.

From these events, we see that P(A/E) is the probability that the item is produced by the plant, given that the item is defective. Also $P(A \cap E)$ is the probability that the items are produced by the plant and are defective. Information in the problem gives the following probabilities in connection with the random selection of a pipe from a day's total production:

(a) Prior probabilities:

$$P(A1) = \frac{500}{500 + 1000 + 2000} = \frac{1}{7}$$

$$P(A2) = \frac{1000}{3500} = \frac{2}{7};\ P(A3) = \frac{2000}{3500} = \frac{4}{7}$$

(b) Likelihoods:

$$P(E/A1) = 0.005;\ P(E/A2) = 0.008;\ P(E/A3) = 0.10$$

(c) Joint Probabilities:

$$P(A1 \cap E) = P(A1)\ P(E/A1) = \left(\frac{1}{7}\right)(0.005) = 0.005/7$$

$$P(A2 \cap E) = P(A2)\ P(E/A2) = \left(\frac{2}{7}\right)(0.008) = 0.016/7$$

$$P(A3 \cap E) = P(A3)\ P(E/A3) = \left(\frac{4}{7}\right)(0.10) = 0.040/7$$

$$P(E) = \Sigma\ P(A1)\ P(E/A1) = 0.061/7$$

(d) Posterior Probabilities:

$$P(A1/E) = \frac{P(A1 \cap E)_1}{P(E)} = \frac{0.005/7}{0.061/7} = \frac{5}{61}$$

$$P(A2/E) = \frac{P(A2 \cap E)}{P(E)} = \frac{0.016/7}{0.061/7} = \frac{16}{61}$$

$$P(A3/E) = \frac{P(A3 \cap E)}{P(E)} = \frac{0.040/7}{0.061/7} = \frac{40}{61}$$

Since P(A2/E) is by far the greatest posterior probability, it is then most probable that the defective item has been drawn from the output of the third plant As a check on the above calculations, the sum of all the posterior probabilities must be unity.

(ii) Probability that the pipe came from the first plant

$$= \frac{\frac{1}{7} \times 0.005}{\left(\frac{1}{7} \times 0.005\right) + \left(\frac{2}{7} \times 0.008\right) + \left(\frac{4}{7} \times .010\right)}$$

$$= \frac{.0007}{.0007 + .0023 + .0057} = \frac{.0007}{.0087} = 0.0805.$$

Example 40:

The odds against A speaking the truth are 4 : 6 while the odds in favour of B speaking the truth are 7 : 3.

(i) What is the probability that A and B contradict each other in stating the same fact ?

(ii) If A and B agree on a statement, what is the probability that this statement is true ?

Solution:

Let us define the events

E1 = A speaks the truth

E2 = B speaks the truth

Then $\overline{E1}$ and $\overline{E2}$ represent the complementary events that A and B tell lie respectively. We are given:

$$P(E1) = \frac{6}{10}\frac{3}{5};\ P\left(\overline{E1}\right) = 1 - \frac{3}{5} = \frac{2}{5}$$

$$P(E2) = \frac{7}{10};\ P\left(\overline{E2}\right) = 1 - \frac{7}{10} = \frac{3}{10}$$

(i) The event E that A and B contradict each other on an identical point can happen in the following mutually exclusive ways:

(a) A speaks the truth and B tells a lie, *i.e.*, the event $E1 \cap \overline{E2}$ happens.

(b) A tells a lie and B speaks the truth, *i.e.*, the event $\overline{E1} \cap E2$ happens.

By applying addition theorem, we get:

$$P(E) = P(E1 \cap \overline{E2}) + P(\overline{E1} \cap E2)$$

$$= P(E1),\ P(\overline{E2}) + P(\overline{E1}).\ P(E2)$$

$$= \left(\frac{3}{5}\times\frac{3}{10}\right)+\left(\frac{2}{5}\times\frac{7}{10}\right)=\frac{23}{50}$$

(ii) If a statement is true, the probability that both A and B agree in asserting that it is true is given by multiplication theorem of probability as:

$$= \frac{3}{5}\times\frac{7}{10}=\frac{21}{50}$$

The probability that both A and B agree in asserting that the statement is false is:

$$\frac{2}{5}\times\frac{3}{10}=\frac{6}{50}$$

Hence the required probability that the statement is true:

$$\frac{21/50}{(21/50)+(6/50)}=\frac{21}{27}$$

Example 41:

A Product is assembled from three components X, Y and Z, the probability of these components being defective is respectively 0.01, 0.02 and 0.05. What is the probability that the assembled product will not be defective?

Solution:

Let A, B and c denote the respective probabilities of components X, Y and Z being defective. We are given

$$P(A) = 0.01, P(B) = 0.02, P(C) = 0.05$$

$$P(A \text{ or } B \text{ or } C) = P(A) + P(B) + P(C) - P(AB) - P(BC) - P(AC) + P(ABC)$$

$$= 0.01 + 0.02 + 0.05 - 0.0002 - 0.0010 - 0.0005 + 0.00001$$

$$= 0.0784.$$

Hence the probability that the assembled product will not be defective

$$= 1 - .07831 = 0.92169 \text{ or } 0.922.$$

Example 42:

An artical manufactured by a company consists of two parts A and B. In the process of manufacture of part A9 out of 100 are likely to be defective. Similarly A out of 100 are likely to be defective in the manufacture of part B. Calculate the probability that the assembled parts will not be defective.

Solution:

P(Part A is defective) = 9/100 = 0.09

P(Part A is not defective) = 1 – 0.09 = 0.91

P(Part B is not defective) = 5/100 = 0.05

P(Part B is not defective) = 1 – .05 = .95

P(Assembled parts will not be defective)

= P(Part A is not defective) × P(Part B is not defective)

= 0.91 × 0.95 = 0.864.

Example 43:

Items produced by a certain process each may have one or both of two types of defects, A and B. It is known that 20 percent of the items have type A defects and 10 percent have type B defects. Furthermore, 6 percent are known to have both types of defects. What is the probability that a randomly selected item will be defective ?

Solution:

We are given that

P(A) = 0.20, P(B) = 0.10, P(AB) = 0.06

∴ Required probability will be given by

P(A or B) = P(A) + P(B) – P(AB) = 0.20 + 0.10 – 0.06 = 0.24

Hence the probability that a randomly selected item will be defective is 0.24.

Example 44:

The probability that a contractor will get a plumbing contract is 2/3 and the probability that he will not get an electric contract is 5/9. If the probability of getting at least one contract is 4/5, what is the probability that he will get both the contracts ?

Solution:

Let A and B denote the events that the contractor will get the plumbing and electric contracts respectively. We are given

$$P(A) = \frac{2}{3}, P(\overline{B}) = \frac{5}{9} \text{ or } P(B) = 1 - P(\overline{B}) = \frac{4}{9}$$

Using addition theorem

P(A or B) = P(A) + P(B) – P(A and B)

or P(A or B) = P(A) + P(B) – P(A and B)

$$= \frac{2}{3} + \frac{4}{9} - \frac{4}{5} = \frac{14}{45} = 0.311$$

Example 45:

The odds against student X solving a Business Statistics problem are 8 to 6, and odds in favour of student Y solving the problem are 14 to 16.

(i) What is the chance that the problem will be solved if they both try independently of each other ?

(ii) What is the probability that none of them is able to solve the problem ?

Solution:

Let A = the event that the first student solves the problem, and

B = the event that the second student solves the problem.

$$P(A) = \frac{6}{8+6} = \frac{6}{14} : P(B) = \frac{14}{14+16} = \frac{14}{30}$$

(i) Probability that the problem will be solved

= P(at least one of them solves the problem)

= P(A or B) = P(A) + P(B) – P(A and B)

= P(A) + P(B) – P(A) × P(B)

$$= \frac{6}{14} + \frac{14}{30} - \frac{6}{14} \times \frac{14}{30} = \frac{73}{105} = 0.695$$

(ii) Probability that neither A nor B solves the problem

$P(\overline{A} \text{ and } \overline{B}) = P(\overline{A}) \times P(\overline{B})$

$$= [1 - P(A)] \times [1 - P(B)] = \frac{8}{14} \times \frac{16}{30} = \frac{32}{105} = 0.305.$$

Example 46:

A company has two plants to manufacture scooters. Plant 1 manufactures 80 percent of the scooters and Plant 2 manufactures 20 percent. In plant 1 only 85 out of 100 scooters are considered of standard quality. In plant 2, only 65 out of 100 scooters are considered of standard quality. What is the probability that a scooter selected at random came from plant 1, if it is known that it is of standard quality ?

Solution:

Let A = The scooter purchased is of standard quality

B = The scooter is of standard quality and came from plant 1

C = The scooter is of standard quality and came from plant 2

D = The scooter came from plant 1.

The percentage of scooters manufactured in plant 1 that are of standard quality is 85 percent of 80 percent *i.e.*, 85 × 80/100 = 68 percent or P(B) = .68.

The percentage of scooters manufactured in plant 2 that are of standard quality is 65 percent of 20 percent, *i.e.*, 65 × 20/100 = 13 percent or P(C) = .13.

The probability that a customer obtains a standard quality scooter from the company is .68 + .13 = .81.

The probability that the scooters selected at random came from plant 1, if it is known that it is of standard quality is given by:

$$P(D/A) = \frac{P(DA)}{P(A)} = \frac{.68}{.81} = .8395 \text{ or } 0.84.$$

Example 47:

In a given business venture a man can make a profit of Rs. 1,000 with probability 0.8 or suffer a loss of Rs. 400 with probability 0.2. Determine his expectation.

Solution:

$$E(X) = P_1X_1 + P_2X_2 + \text{..........} P_kX_k.$$
$$= 0.8 \times 1{,}000 + 0.2 \times (-400) = \text{Rs. } 720.$$

Example 48:

In a bolt factory machines A, B and C manufacture respectively 25%, 35% and 40%. Of the total of their output 5, 4 and 2 percent are defective bolts. A bolt is drawn at random from the product and is found to be defective. What is the probability that it was manufactured by machines A, B and C?

Solution:

Let

A_1 = the event of drawing a bolt produced by First machine.

A_2 = the event of drawing a bolt produced by Second machine.

A_3 = the event of drawing a bolt produced by Third machine.

Then from the first information

$P(A_1) = .25$; $P(A_2) = .35$ $P(A_3) = .40$

From the additional information

B = the event of drawing a defective bolt

$P(B/A_1) = 0.05$; $P(B/A_2) = 0.04$; $P(B/A_3) = 0.02$

Computation of Posterior Probabilities

Event	Prior Probability $P(A_i)$	Conditional Probability $P(B/A_i)$	Joint Probability $P(A_i \times B)$ (2) × (3)	Posterior Probability $P(A_i/B)$ (4) + (B)
(1)	(2)	(3)	(4)	(5)
A_1	0.25	0.05	0.0125	0.0125/0.0345=0.362
A_2	0.35	0.04	0.0140	0.014/0.0345=0.406
A_3	0.40	0.02	0.0080	0.008/0.0345=0.232
	1.00		0.0345	1.000

Without the additional information, we may be inclined to say that the defective item is drawn from machine 3 because $P(A_3) = 0.40$ is larger than $P(A_i) = 0.25$ and $P(A_2) = 0.35$. With the additional information, we may give a better answer. The probability that the item was defective and produced by machine A is 0.362 or 36.2% and by machine B is 0.406 = 40.6% and machine C is 0.232 = 23.2%.

Example 49:

From a computer tally based on employer records the personnel manager of a large manufacturing firm finds that 15 percent of the firm's employees are supervisors and 25 percent of the firm's employees are collage graduates. He also discovers that 5 percent of the firm's employees are both supervisors and collage graduates. Suppose that an employee is selected at random from the firm's personal record, what is the

(i) Probability of selecting a person who is both a college graduate and a supervisor.

(ii) Probability of selecting a person who is neither a supervisor nor a college graduate ?

Solution:

Let A denote the event that the person selected is supervisor and B the event that he is a college graduate. On the basis of given information, we have:

$$P(A) = \frac{15}{1000}; P(B) = \frac{25}{100}; P(AB) = \frac{5}{100}$$

(i) Probability of selecting a person who is both a college graduate and supervisor is:

$$P(A \text{ and } B) = \frac{5}{100} = 0.05$$

(ii) Probability of selecting a person who is neither a supervisor nor a college graduate is:

$$P(\bar{A} \text{ and } \bar{B}) = 1 - P(A \text{ or } B) + 1 - [P(A) + P(B) - P(A \text{ and } B)]$$

$$= 1 - \frac{15}{100} + \frac{25}{100} - \frac{5}{100} = \frac{65}{100} = \frac{13}{20} = 0.65$$

Example 50:

The probability that a trainee will remain with a company is 0.6. The probability that an employee earns more than Rs. 10,000 per year is 0.5. The probability that and employee is a trainee who remained with the company or who earns more than Rs. 10.000 per year is 0.7. What is the probability that an employee earns more than Rs. 10,000 per year given that he is a trainee who stayed with the company ?

Solution:

Let A be the event that a trainee will remain with the company.

Let B be the event that an employee earns more than Rs. 10,000

We are given: P(A) = 0.6, P(B) = 0.5.

P(A management trainee remains with the company or earns more than Rs. 10,000 per year) = 0.7.

$P(A \cup B) = 0.7$

or $P(A) + P(B) - P(A \cap B) = 0.7$

or $P(A \cap B) = P(A) + (B) - 0.7$

Hence the required probability shall be:

$$P(B/A) = \frac{(A \cap B)}{P(A)} = \frac{0.4}{0.6} = 0.667$$

Example 51:

A survey conducted over last 25 years indicated that in 10 years the winter was mild, in 8 years it was cold and in the remaining 7 it was very cold. A company sells 1,000 woollen coats in a mild year, 1,300 in a cold year and 2,000 in a very year.

You are required to find the yearly expected profit of the company if a woollen coat costs Rs. 173 and it is sold to stores for Rs. 248.

Solution:

State of nature (1)	P(x) (2)	Sale of woollen coat (3)	Profit (x) (4)
Mild winter	$\frac{10}{25} = 0.40$	1,000	1,000 (248 – 173)
Cold winter	$\frac{8}{25} = 0.32$	1,300	1,000 (248 – 173)
Very cold winter	$\frac{7}{25} = 0.28$	2,000	2,000 (248 – 173)

Expected Profit:

$$E(x) = (1{,}000 \times 75 \times .4) + (1{,}300 \times 75 \times .32) + (2{,}000 \times 75 \times .28)$$

$$= 30{,}000 + 31{,}200 + 42{,}000 = \text{Rs. } 1,03,200.$$

Example 52:

The Probability that there is at least one error in an accounts statement prepared by A is 0.2 and for B and C they are 0.25 and 0.4 respectively. A, B and C prepared 10, 16 and 20 statements respectively. Find the expected number of correct statement in all.

Solution:

$$E(X) = P_1X_1 + P_2X_2 + P_3X_3$$

$$P_1 = 1 - 0.2 = 0.8;\ P_2 = 1 - 0.25 = 0.75;\ P_3 = 1 - 6.4 = 0.6$$

X_1, X_2 and X_3 = 10, 16 and 20 respectively.

∴ Expected number of correct statements or E(X)

$$= (.8 \times 10) + (.75 \times 16) + (.6 \times 20) = 8 + 12 + 12 = 32.$$

Example 53:

A candidate is selected for interview for 3 posts. For the first post there are 3 candidates, for the second there are 4 and for the third there are 2.

What are the chances of his getting at least one post ?

Solution:

The probability that the candidate gets at least one post

= 1 – probability that the candidate does not get any post

Probability that he does not get the first post $= 1 - \frac{1}{3} = \frac{2}{3}$

Probability that he does not get the second post $= 1 - \frac{1}{4} = \frac{3}{4}$

Probability that he does not get the third post $= 1 - \frac{1}{2} = \frac{1}{2}$

Since the events are independent, probability that he does not get any of the three posts:

$$P(A) \times P(B) \times P(C) = \frac{2}{3} \times \frac{3}{4} \times \frac{1}{2} = \frac{1}{4}$$

Hence the probability that the candidate gets at least one post

$$= 1 - \frac{1}{4} = \frac{3}{4}\ 0.75.$$

Example 54:

Set up a sample space for the single toss of a pair of fair dice. From the sample space, determine the probability that the sum in tossing a pair of dice is either 7 or 11.

Solution:

Sample space for the single toss of a pair of dice.

(1, 6)	(2, 6)	(3, 6)	(4, 6)	(5, 6)	(6, 6)
(1, 5)	(2, 5)	(3, 5)	(4, 5)	(5, 5)	(6, 5)
(1, 4)	(2, 4)	(3, 4)	(4, 4)	(5, 4)	(6, 4)
(1, 3)	(2, 3)	(3, 3)	(4, 3)	(5, 3)	(6, 3)
(1, 2)	(2, 2)	(3, 2)	(4, 2)	(5, 2)	(6, 2)
(1, 1)	(2, 1)	(3, 1)	(4, 1)	(5, 1)	(6, 1).

P(A) = Sum of probabilities associated with each point in A = 6/36.

P(B) = Sum of probabilities associated with each point in B = 2/36.

P(A+B) = Sum of probabilities of points in A, in B or in both

= (6 + 2)/36 = 8/36 = 2/9.

Example 55:

You note that your officer is happy in 60% cases of your calls. You have also noticed that if he is happy, he accedes to your requests with a probability of 0.4, whereas if he is not happy, he accedes to your requests with a probability of 0.1. You call on him one day and he accedes to your request. What is the probability of his being happy ?

Solution:

The probability that he is happy and accedes to requests

$$= .6 \times .4 = .24$$

The probability that he is unhappy and accedes to requests

$$= .4 \times .1 = .04$$

Total probability of acceding to requests

$$= .24 + .04 = 0.28$$

The probability of his being happy if he accedes to requests

$$= \frac{0.24}{0.28} = 0.857.$$

Example 56:

The data for the promotion status and academic qualification regarding 100 employees of a company is an follows :

	MBA	***Academic qualifications Non-MBA***	***Total***
Promotional status			
Promoted	*12*	*48*	*60*
Not promoted	*18*	*22*	*40*
Total	*30*	*70*	*100*

At random one employee is picked up. What is the probability that

(i) *he is an MBA,*

(ii) *he is promoted,*

(iii) *he is promoted given that he is an MBA, and*

(iv) *he is an MBA given that he is promoted ?*

Solution:

Let the various events be defined as follows :

A : an employee is an MBA

B : an employee is promoted

C : an MBA employee is promoted

D : a promoted employee is an MBA

(a) $P(A)=\frac{30}{100}=0.3$

(b) $P(B)=\frac{60}{100}=0.6$

(c) $P(c)=\frac{12}{30}=0.4$

(d) $P(d)=\frac{12}{60}=0.2$

Example 57:

In a certain town, males and females form 50 percent of the population. It is known that 20 percent of the males and 5 percent of the females are unemployed. A research student studying the employment situation selects unemployed persons at random. What is the probability that the person selected is (a) male, (b) female ?

Solution:

Based on the given data the following table depicts the population of the town according to sex and employment status :

	Employed	*Unemployed*	*Total*
Males	0.40	0.10	0.50
Females	0.475	0.025	0.50
Total	0.875	0.125	1.00

Let a male chosen be denoted by M

A female chosen be denoted by F

The male, female chosen is unemployed by U

(a) $P(M/U)=\frac{P(M\cap U)}{P(U)}=\frac{0.10}{0.125}=0.80$

(b) $P(F/U)=\frac{P(F\cap U)}{P(U)}=\frac{0.025}{0.125}=0.20$

Example 58:

A consignment is offered to two firms X and Y for Rs. 1,00,000. The following table shows the probability at which the firms will be able to sell it at different prices :

	Selling Price (Rs.)			
Probability	***80,000***	***90,000***	***1,05,000***	***1,10,000***
X	*0.2*	*0.3*	*0.4*	*0.1*
Y	*0.25*	*0.2*	*0.5*	*0.05*

Which firm X or Y will be more inclined towards the offer ?

Solution:

Firm X

$$E(X) = (80{,}000 \times .2) + (90{,}000 \times .3) + (1{,}05{,}00 \times 0.4) + (1{,}10{,}00 \times .1)$$

$$= 16{,}000 + 27{,}000 + 42{,}000 + 11{,}000 = 96{,}000$$

Firm Y

$$E(Y) = (80{,}000 \times .25) + (90{,}000 \times .2) + (1{,}05{,}000 \times .5) + (1{,}10{,}000 \times .05)$$

$$= 20{,}000 + 18{,}000 + 52{,}500 + 5{,}500 = 96{,}000$$

Since the expectation is the same, both the firms would be equally inclined towards the offer.

Example 59:

The odds against A speaking the truth are 4 : 6 while the odds in favour of B speaking the truth are 7 : 3.

(i) What is the probability that A and B contradict each other in stating the same fact ?

(ii) If A and B agree on a statement, what is the probability that this statement is true ?

Solution:

Let us define the events

E1 = A speaks the truth

E2 = B speaks the truth

Then $\overline{E1}$ and $\overline{E2}$ represent the complementary events that A and B tell lie respectively. We are given:

$$P(E1) = \frac{6}{10}\frac{3}{5}; \; P\left(\overline{E1}\right) = 1 - \frac{3}{5} = \frac{2}{5}$$

$$P(E2) = \frac{7}{10};\ P\left(\overline{E2}\right) = 1 - \frac{7}{10} = \frac{3}{10}$$

(i) The event E that A and B contradict each other on an identical point can happen in the following mutually exclusive ways:

(a) A speaks the truth and B tells a lie, *i.e.*, the event $E1 \cap \overline{E2}$ happens.

(b) A tells a lie and B speaks the truth, *i.e.*, the event $\overline{E1} \cap E2$ happens.
By applying addition theorem, we get:

$$P(E) = P(E1 \cap \overline{E2}) + P(\overline{E1} \cap E2)$$

$$= P(E1), P(\overline{E2}) + P(\overline{E1}). P(E2)$$

$$= \left(\frac{3}{5} \times \frac{3}{10}\right) + \left(\frac{2}{5} \times \frac{7}{10}\right) = \frac{23}{50}$$

(ii) If a statement is true, the probability that both A and B agree in asserting that it is true is given by multiplication theorem of probability as:

$$= \frac{3}{5} \times \frac{7}{10} = \frac{21}{50}$$

The probability that both A and B agree in asserting that the statement is false is:

$$\frac{2}{5} \times \frac{3}{10} = \frac{6}{50}$$

Hence the required probability that the statement is true:

$$\frac{21/50}{(21/50) + (6/50)} = \frac{21}{27}$$

EXERCISES

1. (a) In a random sample of 1989 wheat fields, 1,124 were irrigated, 568 grew wheat mixed with barley and 208 were both irrigated and mixed with barley. What is the probability that a field selected at random will be both irrigated and pure?

   [1/4]

   (b) A factory finds that on an average 20% of the bolts produced by a given machine will be defective for certain specified requirement. If 10 bolts are selected at random from the day's production of this machine, find the probability tat (i) exactly 2, (ii) 2 or more will be defective.

   [(i) 0.302, (ii) 0.6242, (iii) 0.0063]

2. (a) What is the probability that a leap year selected at random will contain 53 Tuesdays?

   (b) An umbrella salesman can earn Rs. 400 per day in case of rain but will lose Rs. 100 per day if it does not rain Find EMV, if probability of rain is 0.4.

   [Rs. 100].

3. (a) In a class of 75 students, 15 were considered to be very intelligent, 45 as medium and the rest below average. The probability that a very intelligent student fails in a viva voice examination is 0.05, the medium student failing has a probability 0.05 and the corresponding probability for a below average student is 0.15. If a student is known to have passed the viva voice examination, what is the probability that he is below average?

   [0.18]

   (b) A local politician claims that the assessed value of houses for property tax by M.C.D. is not correct in 90% cases. Assuming that the claim is true, what is the probability that out of a sample of 4 houses selected at random :

   (i) at least one will be found to be wrongly assessed.

   (ii) at least one will be found to be correctly assessed.

   [(i) 0.9999 (ii) 0.344]

4. (a) On an average 20% of persons going to a handicraft emporium are foreigners and the remaining 80% are local persons, 75% of such foreigners and 50% of such local persons are found to make purchases. If a bundle of purchased items is sent to the cash counter, what is the probability that the purchaser is a foreigner?

   (b) A bag contains 5 white and 3 black balls. Two balls are dawn at random one after the other without replacement. Find the probability that balls drawn are black.

5. (a) Urn 1 contains 2 white and 3 black balls and urn ll contains 3 white and 4 black balls. One ball is transferred from the first urn to the second urn and thereafter one balls is transferred from the second urn to the first urn. What is the probability that the ball drawn will be white?

   (b) Three newspapers A, B and c are published in a certain city. It is estimated from a survey of the adult population 20% read A,

16% read B, 14% read C, 8% read both A and B, 5% read both A and C, 4%, read both B and C, 2% read all the three. What percentage reads at least one, and what percentage reads both A and B?

6. (a) If it rains, a dealer of raincoats can earn Rs. 400 per day. If it is fair he can lose Rs. 80 per day. What is his expectation if the probability of a fair day is 0.6?

   (b) Two balls are drawn from a bag containing 8 red and 7 white balls. Find the chance that :

   (i) they are both red.

   (ii) they are both white, or

   (iii) One is red and the other white.

   [(i) 4.15, (ii) 1/5, (iii) 4/15]

   (c) From the 20 tickets marked with the first 30 natural numbers, one is drawn at random. Find the chance that the number on it is multiple of 1 or 5.

   [7/15]

7. Two groups each of three children contain respectively two boys and one girl and one boy and two girls. One child is drawn at random from each group. Calculate the probability that (a) both will be boys. (b) one boy and the other girl, and (c) at most one boy will be selected.

   [(a) 2/9, (b) 5/9, (c) 7/9]

8. (a) During war one ship in 10 was sunk on the average in making a certain voyage. Find the probability that at least 3 out of a convoy of 6 ships would arrive safely.

   (b) Three houses of the same type were advertised to be let in a locality. Three men made separate applications for a house. What is the probability:

   (i) that all the three made applications for the same house:

   (ii) that each of the three applied for different houses: and

   (iii) that two of item applied for the same house and the third for one of the other houses?

   [(a) 9,98730/10,00,000, (b) (i) 1/9, (ii) 2/9, (iii) 2/3]

9. (a) In an examinations 30% of the students have failed in Mathematics, 20% of the students have failed in Chemistry and 10% have failed in both Mathematics and Chemistry. A student is selected at random.

(i) What is the probability that the student has failed either in Mathematics or in Chemistry?

(ii) What is the probability that the student has failed either in Mathematics or in Chemistry?

[(i) 1/2; (ii) 0.40]

(b) Two sets of candidates are competing for the positions on the Board of Directors of a company. The probabilities that the first and second sets will win are 0.6 and 0.4 respectively. If the first set wins, the probability of introducing a new product is 0.8, and the corresponding probability, if the second set wins, is 0.4. What is the probability that the new product will be introduced.

10. The following table gives a distribution of wages :

Weekly wages (Rs.) :							
30-35	35-40	40-45	45-50	50-55	55-60	60-65	65-70
No. of wage-earners :							
9	108	488	230	112	30	16	7

An individual is taken at random from the above group; find the probability that :

(a) his wages were under 40.

(b) his wages were 55 or over.

(c) his wages were either between 45-50 or 30-40.

11. (a) the probabilities that the present fiscal and monetary policies will remain unchanged after 20 years from now, are 0.9 and 0.8 respectively. Find the probability that in 20 years : (i) both, (ii) neither, (iii) at least one of the two policies will remain unchanged.

(b) Four persons in a group of 20 are graduates. If 4 persons are selected at random from 20, find the probability that :

(i) all are graduates

(ii) at least one is a graduate.

12. (a) A bag contains 5 white and 4 black balls. One ball is drawn from the bag and replaced and then a second draw of a ball is made. What is the chance that the two balls drawn are of different colours?

(b) Out of 320 families with 5 children each, what percentage would be expected to have (i) 2 boys, and 3 girls (ii) at least one boy? Assume equal probability for boys and girls.

13. (a) Explain what do you understand by the term 'probability'. State and prove the addition and multiplication theorems of probability.
    (b) Discuss the various definitions of probability explaining the terms used.
14. Explain the terms 'mutually exclusive' and 'equally likely'.

    In terms of probability, explain an independent event. Is it possible that two events may be independent but not mutually exclusive? Give example.
15. State what do you understand by the term 'mathematical expectation'. How is it useful for a businessman? Give an example to illustrate its usefulness.
16. Explain the terms mutually exclusive events and independent events. Give one example for each. State and prove the addition rule of probability.
17. Explain with examples the concepts of independent and mutually exclusive events in probability. State and prove the multiplicative theorem of probability? What is its modified form when the events are independent?
18. (a) Explain with examples the rules of addition and multiplication in theory of probability.
    (b) Define with examples mutually exclusive events. Mutually independent events and exhaustive events.
    (c) Explain the term probability and give its multiplicative law.
19. (a) Give the classical definition of the probability explaining the various terms used.
    (b) State and prove additive law of probability. State the multiplicative law also.
20. (a) Explain the concepts of additive, multiplicative and conditional probability with suitable samples.
    (b) Point out the difference between objective and subjective probabilities.
    (c) Explain the two theorems of probability.
21. (a) Define conditional probability, Explain multiplication theorem for the dependent events.
    (b) Explain either the joint and conditional probabilities or the concept of posterior probabilities.

22. (a) Develop the additive probability role for (a) Mutually exclusive events, (b) Overlapping events, and (c) Complementary events.
    (b) Distinguish between 'a priori' and 'a posteriori' probability.
23. (a) State and prove the addition and multiplication theorems of probability giving one example of each rule.
    (b) Define the terms (i) Random experiment, (ii) Sample space, and (iii) event.
    (c) State and prove Bayes 'Theorem of probability.
24. (a) State the general addition rule for probability. What is its form if the concerned events are mutually exclusive?
    (b) Give the three definitions of probability, stating when you will use each of them.
25. (a) Critically evaluate the classical definition of probability.
    (b) Develop the additive rule of probability for independent and dependent events.
26. (a) Define 'Probability' and explain briefly the importance of this concept in managerial decision-making.
    (b) Give axiomatic and relative frequency definition probability. State the addition rule of probability. Define conditional probability.
27. (a) State and explain Bayes' theorem and bring out its importance in probability theory.
    (b) Explain the various approaches to probability. Are these approaches contradictory?
28. Explain the various approaches to probability. Are these approaches contradictory?

    What is conditional probability? Explain with the help of an example.
29. Explain with examples the various school of thought on probability. Discuss the importance of probability in business decision-making.
30. (a) Explain the concept of probability and point out its role in business decision-making.
    (b) What is conditional probability? Explain with the help of an example. Discuss the importance of probability in statistics.
31. (a) Define the concepts of conditional probability and independent events.
    (b) Explain what do you understand by the term probability. Discuss its importance in decision-making.

32. (a) Explain the concepts of simple, joint, conditional and marginal probabilities. Give examples to illustrate these probabilities.

    (b) State and explain Baye's theorem and bring out its importance in probability, theory.

    (c) State the relationship between ^nP_r and ^nC_r.

    (d) In how many ways can the word 'management' be arranged?

33. (a) Distinguish between conditional probability and joint probability under conditions of statistical dependence with help of an example.

    (b) What do you mean by empirical approach to probability? Discuss the importance of probability in statistics.

34. (a) State and prove the multiplicative theorem of probability. How is the result modified when the events are independent?

    (b) State and prove the addition and multiplication theorems of probability for two mutually exclusive events.

35. (a) State the addition and multiplication theorems of probability and give two different examples illustrating the application of these theorems.

    (b) Define independent and mutually exclusive events. Can two events be mutually exclusive and dependent simultaneously? Support your answer with an example.

36. (a) Explain with examples the difference between 'independent' and 'mutually exclusive' events in the theorem of probability. State the theorem of total and compound probabilities.

    (b) State the addition theorem of probability for two events : (i) when they are mutually exclusive, and (ii) when they are not mutually exclusive.

37. (a) Examine critically the different schools of thoughts in probability.

    (b) Distinguish clearly between probability and non-probability sampling methods. Explain any three methods of sampling known to you highlighting their merits and demerits.

    (c) Distinguish clearly the difference between any four of the following concepts:

    (i) Mathematical and statistical probability.

    (ii) Sample and compound events.

    (iii) Independent and dependent events.

(iv) Mutually exclusive and independent events.

(v) Permutation and combination.

(d) State and prove the theorem of compound probability. If the events are independent, what will be the form of the theorem.

38. (a) What do you understand by the term mathematical expectation. How is it useful to businessmen? Give one example to illustrate its use.

(b) Prove the addition theorem of probability.

39. (a) Distinguish between any four of them:

(i) Simple and compound events.

(ii) Dependent and independent events.

(iii) Mutually exclusive and compound events.

(iv) Mutually exclusive and independent events.

(v) Priori and posteriori probabilities.

(b) Define conditional probability. State and prove addition theorem on probability.

(c) Explain with examples the various school of thought on probability.

(d) Distinguish between objective and subjective probabilities. Cite an example of each type of probability.

(e) Prove the multiplication theorem of probability.

40. (a) Two cards are drawn from a well-shuffled pack of 52 cards. Find the probability that they are both aces if the first is (i) replaced, and (ii) not replaced.

[(i) 1/16; (ii) 1/221]

(b) Urn A contains 5 red balls and 5 black balls, urn B contains 4 red and 8 black balls and urn C contains 3 red and 6 black balls. A ball is drawn from A, colour unknown, and put into B. Then a ball is drawn from B, colour unknown, and put into C. What is the probability that ball now drawn from C will be red?

[87/260]

41. (a) What is the probability that a vowel selected at random in an English book is an i?

(b) In a single throw with two dice find the chance of throwing (i) 8, and (ii) 11.

[(i) 5/36, (ii) 1/81]

42. (a) Three perfect coins are tossed together. What is the probability of getting at least one head?

[7/8]

(b) A can hit a target 3 times in 5 shots, B 2 times in 5 shots and C3 times in 4 shots. They fire a volley. What is the probability of hitting 2 shots?

43. The personnel department of a company has records which show the following analysis of its 200 engineers:

Age	*Bachelor's degree*	*Master's Degree*	*Total*
Under 30	90	10	100
30 to 4	20	30	50
Over 40	4	10	50
Total 150	50	200	

If one engineer is selected at random from the company, find :

(a) the probability that he has only bachelor's degree.

(b) The probability that he has master degree given that he is over 40.

(c) The probability that he is under 30, given that he has only a bachelor's degree.

44. (a) In a fruit shop, 50 per cent of the fruits are mangoes, 30 per cent oranges and 20 per cent apples, 2 per cent of the mangoes, 1 per cent of the oranges and 0.5 per cent of the apples are bad. One of the three different fruits was taken at random from the shop and was found to be bad. What is the probability that the fruit taken was and orange?

(b) A pizza shop found that 75% of all customers use tomato sauce, 80% use a special sweet sour preparation called 'Punch' and 65% use both when ordering a pizza. What is the probability that a customer will use at least one of these?

[0.90]

45. (a) A bag contains 10 balls of which 4 are black and 6 are red. Four balls are drawn. Find the probability of drawing exactly two red balls.

[3/73]

(b) A dice is thrown three times. What is the chance that on the first throw it falls with number upwards, and in the second with either number 1 or number 2, and in the third with either number 1 or 2 or 3 upwards?

46. (a) Answer the following question on probability :

10% of the employees of a certain company have been to public schools. Of these, 70% hold administrative positions. Of these that have not been to public school, 30% hold administrative positions. If an employee is selected at random from the administrative staff, what is the probability that he was educated in a public school?

[7/34]

(b) A person applies for the post of manager in two firms A and B. He estimates that the probability of his being selected in firm A is 0.72, being rejected in firm B is 0.45 and being rejected in at lest one of the firms is 0.55. Find the probability that he will be selected in at least one of the firms.

47. The probability that India wins a cricket Test match against Pakistan is given to be 1/3. If India and Pakistan play three Test matches, what is the probability that:

(i) India will lose all three Test matches, and

(ii) India will win at least one Test match?

[(i) 8/27, (ii) 19/27]

48. (a) In a throw of two dice, what is the probability of getting:

(i) A total of 7 or 11 of the digits on the face upward.

(ii) The total would be more that 5 of the digits on the face upward?

[2/9; 13/18]

(b) (i) What is the probability of drawing a king from a pack of playing cards?

(ii) A shopkeeper has surplus stock of 9 different kinds of goods. How many different parcels may he produce?

(iii) From a set of 19 cards numbered 1,2,3,.......19, one is drawn at random. Find the chance that its number is divisible by 3 or 7.

(iv) In how many ways the letters of word "*STATISTICALLY*" can be arranged?

[(i) 1/13, (ii) 5/11, (iii) 8/19, (iv) 64864800]

49. (a) A problem of Statistics is given to three students for solution. Their probabilities of solving it are 1/2, 1/3 and 1/4 respectively, What is the probability that the problem will be solved?

(b) A group consists of 100 men and 80 women, 40 among men and 50 among women are graduates. If one person is selected at random from the group, find the probability the person is either a woman or a graduate.

50. (a) Define probability and explain the importance of this concept in Statuses.

(b) Explain the various approaches to probability. Are these approaches contradictory?

(c) Define probability. Briefly explain the different schools of thoughts on probability.

(d) When are two events said to be independent in the probability sense? Give examples of dependent and independent events.

51. (a) (i) Explain the theorems of probability.

(ii) What is joint probability?

(iii) What is inverse probability?

52. (a) Define mutually exclusive events. State and prove addition theorem of probability for two events.

(b) Define probability. Does probability always relate to one event only ? Give examples.

(c) Examine critically the different schools of thought on probability.

53. (a) A player tosses four fair coins. He wins Rs. 16 if four tails occur; Rs. 8 if three tails occur; Rs. 4 if two tails occur and Rs. 2 if one tail occurs. If the game is to be fair how much should he win or lose in case no tail occurs?

(b) There are 100 students in a college class of which 36 are boys studying Statistics and 13 are girls not studying Statistics. If there are 55 girls in all, find the probability that a boy picked up at random is not studying Statistics.

54. (a) The odds again student X solving Business Statistics problem are 8:6 and odds in favour of student Y solving the sample problem are 14:16:

(i) What is the chance that the problem will be solved if they both try?

(ii) What is the probability that they both working independently of each other, solve the problem?

(iii) What is the probability that neither solves the problem?

(b) In a colony, 5,000 persons are residing, out of which 1,200 are above 30 years of age and 3,000 are females. Out of 1,200 who are above 30 years, 200 are females. Suppose after a person is chosen you are told that the person is a female. What is the probability that 'she is above 30 years of age?

55. (a) Five coins whose faces are marked 2 and 3 are thrown. What is the probability of obtaining a total of 12?

(b) If it rains, a taxi driver can earn Rs. 100 per day. If it is fair, he can lose Rs. 10 per day. If the probability of rain is 0.4, what is his expectation?

56. (a) A husband and wife appear for an interview for two vacancies in the same post. The probability of husbands selection is 1/3 and that of wife's selection is 1/5. Find the probability that :

(i) both of them will be selected.

(ii) only one of them will be selected.

(b) A company has 300 employees of whom 20 are men. When questioned, 180 people agreed that they were happy in their work. What is the probability that a man employee in the company is unhappy?

(c) A bag contains 6 white, 4 red and 10 black balls. Two balls are drawn at random. Find the probability that they will both be black.

(d) In a class of 50 students, 20 play football and 16 play basket ball. If 10 play both the games, find out how many play neither.

57. (a) What are mutually exclusive events? State which of the following are mutually exclusive .

(i) A : a red card, B : an ace in a draw of a card from a pack.

(ii) A : a total of 7, B: an odd number on each die in a simultaneous throw of 2 dice. Give reasons for you answer.

(b) A manufacturing firm produces T.V. sets in three plants with daily production value of 250, 500 and 1,000 units respectively. According to past experience, it is known that the fractions of defective output produced by the three plants are respectively, .005, .008 and .010. If a T.V. set is selected from a day's total production and found to be defective, find out (i) from which plant the T.V. comes, (ii) what is the probability that it comes from the second plant?

58. (a) A bag contains 8 red and 7 white balls. Two balls are drawn from the bag. Find the probability that: (i) they are both red (ii) they are both white (iii) one red and the other white.

    (b) In the play of two dice, the thrower loses if the first throw is 2, 4 or 12. He wins if the first throw is a 5 or 11. Find the ratio between his probability of losing and probability of winning in the first throw.

59. (a) A speaks truth in 60 percent cases and B speaks truth in 70 percent cases. In what percentage of cases are they likely to contradict each other in stating the same fact?

    (b) A speaks truth is 80% cases and B in 90% of the cases. In what percentage of cases they are likely to contradict each other in stating the same fact?

60. (a) A can solve a problem of statistics in 4 out of 5 chances and P can do it in 2 out of 3 chances. If both A and B try to solve the problem, find the probability that it will be solved.

    (b) (i) In how many ways can be alphabets of the word *LOOPHOLE* be arranged?

    (ii) In throw of 2 rectangular dice containing numbers from 1 to 6 on their 6 sides, what are the probabilities of getting (a) a total of 12. and (b) a total of 10.

    (iii) The probability of a cricket team winning match at Kanpur is 2/5 and losing match at Delhi is 1/7. What is the probability of the team winning at least one match?

    (iv) The probability of selling a tin of ghee of four different trademark in a shop are given below:

Trade Mark :	*A*	*B*	*C*	*D*
Probability of a tin being sold :	0.24	0.30	0.50	0.40

    A buyer purchased a tin of ghee from that shop. What is the probability that it was of 'A' trade mark?

    (v) A and B enter into a bet according to which A will get Rs. 200 if it rains on that day and will lose Rs. 100 if it does not rain. The probability of raining on that day is 0.7. What is the mathematical expectation of A?

61. (a) A company is planning to market a new product A. If its competitor does not market a similar product, the probability that product A will do well is 0.8. If the competitor also markets

similar product, the probability that product A will do well is 0.4. It is believed that the probability that the competitor will market a similar product is 0.6. What is the probability that product A will do well if it is marketed?

[0.56]

(b) The probabilities of the students A, B, C solving a problem are respectively 1/2, 1/3, 1/4. What is the probability that the problem is solved if they attempt to solve it independently?

[3/4]

62. Tick the correct answer:

(a) Much of the development in the theory of probability is associated with the name of (i) Fisher, (ii) Karl Pearson (iii) Gosset (iv) Bayes (v) none of these.

(b) The classical school of thought on probability assumes that all possible outcomes of an experiment are:

(i) equally likely (ii) independent, (iii) mutually exclusive, (iv) dependent, (v) mutually exclusive and equally likely.

(c) If an event cannot take place, the probability will be (i) +1, (iii) 0, (iv) none of these

(d) $p \times q$ would always be (i) less than one, (ii) more than one, (iii) zero, (iv) would vary between $\pm$ 1. (v) one.

(e) Additional theorem states that if two events A and B are mutually exclusive, the probability of occurrence of either A and B is given by (i) P (A) + P (B), (ii) P (B) × P (A), (iii) P (A) - P (B), (iv) P (A) × P (B) + P (AB), (v) none of these.

(f) If two events A and B are independent, the conditional probability that they will both occur is given by

(i) P (A) + P(B), (ii) P (A) × P(B), (iii) P(A) - P(B), (iv) P(A) × P(L) + P(AB), (v) P(A) + P(B) - P(AB).

(g) If two events A and B are dependent, the conditional probability of B given A. *i.e.* P(AB) is calculated as :

(i) P(AB) (AB), (ii) P(A)/(B), (iii) P(AB)/P(A) (iv) P(A)/P(B) (v) P(B)/(AB)

(h) If two events A and B are dependent, the conditional probability of A given B, *i.e.*, P(A/B) is calculated as :

(i) P (B/A) (AB), (ii) P (B)/P (A), (iii) P (AB)/P(A), (iv) P (AB)/ P(B), (v) none of these.

(i) The expectation of X is defined as :

(i) p1 X1+ p2 X2 + ... + Pk Xk

(ii) p1 X1 p2 X2 .... Pk Xk

(iii) p1 X2+ p2 X2 + Pk Xk

(iv) p1 X1× p2 X2 + ... + Pk Xk

(v) none of these.

(j) C_2 is equal to

(i) 20, (ii) 10, (iii) 30, (iv) 100, (v) 0.

(k) If P (AB) is equal to 0.24 and P (A) is equal to 0.60, then P(A/B) is ....

(i) 0.16, (ii) 0.36, (iii) 0.84, (iv) none of these

[Ans. : a (iv), b (v), c (iii), d (v), e (i), f (ii), g (iii), h (iv), i (i), j (ii), k (iii)]

63. (a) The probability that X passes Mathematics is 2/3, the probability that he passes Statistics is 4/9. If the probability of passing at least one subject is 4/5, what is probability that Mr. X would pass both the subjects?

[14/45]

(b) The records of 400 examiners are give below :

	Educational qualification			
Score	B.A.	B.Sc.	B. Com.	Total
Below 50	90	30	60	180
Between 50 & 60	20	70	70	160
Above 60	10	30	20	60
Total	120	130	150	400

If an examinee is selected from this group of examinees, find

(i) The probability that he is a commerce graduate;

(ii) The probability that he is a science graduate, given that his score is above 60; and

(iii) The probability that his score is below 50, given that he is a B.A.

[(i) 3/8; (ii) 3/13; (iii) 3/4]

64. (a) Each of the three identical jewellery boxes has 2 drawers. In each drawer of the first box there is a gold watch. In each drawer of the second box there is a silver watch. In one drawer of the

third box there is a gold watch while in the other drawer there is a silver watch. If we select a box at random, open one of the drawers and find it to contain a silver watch, what is the probability that the other drawer has the gold watch?

(b) A and B throw one dice for a prize of Rs. 11 which is to be won by the player who first throws 6. If A has the first throw what are their respective expectations?

(c) A player tosses three fair coins. He wins Rs. 12 if three tails occur, Rs. 7 if two tails occur and Rs. 2 if only one tail occurs. If the game is to be fair, how much should he win or lose in case no tail occurs?

65. (a) Suppose it is 9 : 7 against a person A, who is now 35 years of age living till he is 95, and 3:2 against a person B, now 45 years of age living till he is 75. Find the chance that at least one of these persons will be alive 30 years hence.

(b) A bag contains 10 red, 5 white and 4 blue balls. If 4 balls are drawn at random, determine the probability that - (i) all 4 are blue balls, and (ii) 1 red and 2 are white.

66. (a) If the probability that the value of certain stock will remain the same is 0.46, the probabilities that its value will increase by 50 paise or Re. 1.00 per share are respectively 0.17 and 0.23 and the probability that its value will decrease by Re. 0.25 per share is 0.14, what is the expected gain per share?

(b) From a sales force of 150 people, one will be chosen to attend a special sales meeting. If 52 are single and 72 are college graduates, and 3/4 of 52 that are single are college graduates, what is the probability that a sales person selected at random will be neither single nor a college graduate?

[13, 130]

67. (a) Two computers A and B are to be marketed. A salesman, who is assigned the job of finding customers for them, has 60% and 40% chances respectively of succeeding in case of computers A and B. The two computers can be sold independently. Given that he was able to sell at least one computer, what is the probability that computer A has been sold?

(b) State and explain Bayes' Theorem. From a pack of cards, two cards are drawn, the first being replaced before the second is drawn, what is the probability that first is a heart and second is not a king?

68\. (a) The probability that India wins a cricket Test match against England is given to be 1/3. If India and England play three Test matches, what is the probability that:

(i) India will lose all the three Test matches?

(ii) India will win at least one Test match?

[(i) 8/27; (ii) 19/27]

(b) The probabilities of X, Y and Z becoming managers are 4/9/, 2/9 and 1/3 respectively. The probabilities that the Bonus scheme will be introduced if X, Y and Z become managers, are 3/10, 1/2 and 4/5 respectively, (i) What is the probability that the Bonus scheme will be introduced, (ii) what is the probability that the manager appointed was X?

[(i) 23/45; (ii) 6/23]

69\. (a) Three contractors A, B and C are bidding for the construction of a new cinema hall. An expert in industry believes that A has exactly half the chance that B has; B, in turn, is 4/5 as likely as C wins the contract. What is the probability for each to win the contract if the expert's estimates are accurate?

[A2/ 11, B4/11, C5/11]

(b) Radio values are tested by subjecting them to a number of electric shocks. Each shock has an independent chance 0.8 of destroying the valves. How many shocks must be given to the valve in order that the probability of the valve being destroyed is 0.99?

[3 shocks]

70\. A factory produces a certain type of output by three machines. The respective daily production figures are:

Machine	'X'	2,000	Units
"	'Y'	3,000	"
"	'Z'	5,000	"

Past experience shows that 105 per cent of the output produced by machine 'Y' and 1.8 per cent of the output by machine 'Z' are defective. An item is drawn at random from the day's production run and is found to be defective. What is the probability that it comes from the outputs of (i) machine 'X', (ii) machine 'Y', and (iii) machine 'Z'?

71\. An Auditor has in his file 12 accounts of which four contain procedural error in posting account balances.

(a) If the auditor randomly selects two of these accounts without replacement, what is the probability that neither account will contain the procedural error?

(b) If the auditor samples three accounts, what is the probability that none of the accounts includes procedural error?

(c) If the auditor samples one account randomly, what is the probability that it will contain the error?

(d) If the auditor samples two accounts randomly, what is the probability that at least one of them will contain the error?

(e) If the auditor samples three accounts randomly, what is the probability that at least one will contain the error?

72\. (a) A researcher has to consult a recently published book. The probability of its being available is 0.5 for Library A and 0.7 for Library B. Assuming the two events to be statistically independent, find out the probability of the book being available in Library A and not available in Library B.

[0.5]

(b) In a large group of students, 80 percent have recommended Statistics book. 3 students are selected at random. (i) Find the probability distribution of the number of students having the Book (ii) Calculate the mean and variance of the Distribution.

73\. (a) An investment firm purchases three stocks for one week trading purposes. It assesses the probability that the stocks will increase in value over the week as 0.8, 0.75 and 0.69 respectively. What is the probability that all the three stocks will increase, assuming that the movements of these stocks are independent? Is this a reasonable assumption?

(b) Two events A and B are mutually exclusive:

P (A) = 1/5 and P (B) = 1/3, find the probability that

(i) either A or B will occur, (ii) both A and B will occur, (iii) neither A nor B will occur. (B. Com., Shivaji Univ., 1998)

[(i) 8/15, (ii) 0, (iii) 7/15]

74\. (a) The probability that a man will be believe for 25 years is 3/5 and the probability that his wife will be alive for 25 years is r/3. Find the probability that (i) both will be alive (ii) only the man will be alive (iii) only the wife will be alive (iv) at lest one will be alive.

[(i) 2/5, (ii) 1/5, (iii) 4/15, (iv) 3/4]

(b) A manager has two assistants and he bases his decision on information supplied independently by each of them. The probability that he makes a mistake in his thinking is 0.005. The probability that an assistant gives wrong information is 0.3. Assuming that the mistakes made by the manager are independent of the information give by the assistants, find the probability that he reaches a wrong decision.
[0.513]

75. (a) Out of 5 vowels and 7 consonants, how many words can be made taking 3 vowels and 3 consonants?

(b) A problem in Statistics is given to the three students P, Q and R whose chances of solving it are 1/3, 1/4 and 2/5 respectively. What is the probability that the problem will be solved?

76. (a) A card is drawn at random from a full pack of 52 playing cards. What is the probability that the drawn card will be either a black card or an Ace?

(b) Find the probability of drawing a queen, a king and jockey in that order from a pack of cards in three consecutive draws, the cards drawn not being replaced.

(c) A person is known to hit the larger in 3 out of 4 shots, whereas another person is known to his the larger in 2 out of 3 shots. Find the probability of the target being his when both of them try.

77. (a) If a player plays a game of chance where he can win Rs. 1,000 with probability 0.5, win Rs, 500 with probability 0.3 and lose Rs. 3,000 with probability 0.2, what is his expected` gain in one play of the game?

(b) Explain whether or not each of the following claims could be correct:

(i) A businessman claims the probability that he will get contract A is 0.5 and that he will get contract B is 0.20. Furthermore, he claims that the probability of getting A or B is 0.50

(ii) A market analyst claims that the probability of selling ten million pounds of plastic A or five million pounds of plastic B is 0.60. He also claims that the probability of selling ten million pounds of A and five million pounds of B is 0.45.

[(i)–0.15 (ii) Since P=1.05, claim is wrong]

78. Two computers, say, A and B, are to be marketed, A salesman, who is assigned the job of finding customers for them, has 60% and 40%

chances respectively of succeeding in case of computers A and B. The computers can be sold independently. Given that he was able to sell at least one computer, what is the probability that A has been sold?

[0.79]

79. Out of 800 families with 4 children each, what percentage would be expected to have (a) 2 boys and 2 girls, (b) at least one boy, (c) no girls, and (d) at most 2 girls. Assume equal probability for boys and girls?

[(a) 37% (b) 93.75% (c) 6.15% (d) 68.75%]

80. (a) The probability that a boy will get a scholarship is 0.90 and that a girl will get 0.80. What is the probability that at least one of them will get the scholarship?

[(a) 5/2: (b) 0.80]

(b) If a man purchases a raftle ticket he can win a first prize of Rs. 5,000 or a second prize of Rs. 2,300 with probabilities 0,001 and 0.003 respectively. What should be a fair price to pay for the ticket?

[11.907]

81. (a) In a certain recruitment test there are multiple choice questions. There are 4 possible answers to each question and of which one is correct. An intelligent student knows 90 per cent of the answers while a weak student knows only 20 per cent.

(i) If an intelligent students gets the correct answer, what is the probability that he was guessing?

(ii) If a weak student gets the correct answer, what is the probability that he was guessing?

[(i) 0,025/0.925; (ii) 0.020/0.220]

(b) The compressors used in refrigerators are manufactured by XYZ Co. at three factories at Pune. Nasik and Nagpur. It is known one which produces the same numbers as the Nagpur one (during the same period), experience also shows that 0.2% of the compressors produced at Pune and Nasik are defective and so are 0.4% of those produced at nagpur.

A quality control engineer, while maintaining a refrigerator, finds a defective compressor. What is the probability that the Nasik factory is not to be blamed?

82. A coin is tossed. If it turns up heads, two balls will be drawn from urn A, otherwise two balls will be drawn from urn B. Urn A contains

three black and one white balls. In both cases, selections are to be made with replacement. What is the probability that urn A is used give that both the balls drawn are black?
[144/1.619].

83. Which of the following statements are true or false.
    (a) The classical approach to probability is the oldest and simplest. T/F
    (b) The probability of throwing eight with a single dice is 1/6. T/F
    (c) The modern probability theory has been developed axiomatically in which probability is an unrefined concept. T/F
    (d) In most fields of research, a prior probability is employed. T/F
    (e) Dependent events are those in which the outcome of one does not affect and is not affected by the other. T/F
    (f) Probability derived from past experience is called empirical probability . T/F
    (g) The probability of obtaining a 3 and 4 in a throw of two dice is 2/30. T/F
    (h) The conditional probability of the given A is written as P(A/B). T/F
    (i) Bayes' Theorem provides us with another formula for computing conditional probability. P(A/B). T/F
    (j) If two events are not mutually exclusive, they must be independent events. T/F
    (k) A statistician who uses objective probability is called a Bayesian statistician. T/F

    [**Ans.** : (a) T (b) F (c) T (d) F (e) F (f) T (g) T (h) F (i) T (j) F (k) F]

84. Fill in the blanks:
    (a) Probability ranges from .... to..................
    (b) The probability obtained by the following relative frequency definition is called ....probability.
    (c) Two events are said to be disjoint when .... simultaneously..........
    (d) If A and B are mutually exclusive events. P (AB)=.....................

(e) The probability that the throw of two dice yields total of 5 is ....

(f) In case of conditional probability .... and B............

(g) Probability theory had its origin in .... games.

(h) (i) ${}_5C_2$= .... (ii) ${}^5p_3$= ....

(i) The order of arrangement is important in ....

(j) The personalistic view of probability regards probability as a measure of ....in a particular event or proposition.

(k) If a card is drawn from a pack of cards. the probability of getting either a king or queen is ....

(l) Joint probability is the probability of the ....occurrence of two or more events.

[**Ans.** : (a) 0 to 1. (b) a posteriori, (c) both cannot happen, in a single trial, (d) zero, (e) 1/2, (f) P (B)′P (A/B) or P(A)′ P (B/A), (g) Gambling, (h) (i) 10, (ii) 60, (i) Permutation, (j) Personal Belief, (k) 2/13, (l) Joint or simultaneous.

2

Theoretical Distributions

INTRODUCTION

The probability distribution of a random variable may be:

1. An empirical listing of outcomes and their observed relative frequencies.
2. A subjective listing of outcomes associated, with their subjective or 'contrived' probabilities representing the degree of conviction of the decisionmaker as to the likelihood of the possible outcomes.
3. Theoretical listing of outcomes and probabilities which can be obtained from a mathematical model representing some phenomenon of interest.

A probability distribution for a discrete random variable is a mutually exclusive listing of all possible numerical outcomes for that random variable such that a particular probability of occurrence is associated with each outcome.

In case of a fair six-faced dice that will not stand on edge or roll out of sight (null events) the probability distribution for the outcomes of single roll of the dice shall be as follows:

PROBABILITY DISTRIBUTION OF THE RESULTS OF ROLLING ONE FAIR DICE

Face of outcome	*Probability*
1	1/6
2	1/6
3	1/6
4	1/6
5	1/6
6	1/6

Since all possible outcomes are included, this listing complete (or collectively exhaustive) and thus the probabilities must sum up to 1.

From the above table we can obtain various probabilities for the rolling of a fair dice. For example.

The probability of a face [: :] = 1/6.

The probability of an even face using the addition rule for mutually exclusive events is :

P(even) = P ([· ·]) + P ([: :]) + P ([: : :])

= 1/6 + 1/6 + 1/6 = 3/6

The probability of a face of [· · ·] or less is

P([· · ·] or less) = P([·]) + P([· ·]) + P ([· · ·])

= 1/6 + 1/6 + 1/6 = 3/6

And the probability of a face greater than [: : :] is

P(> [: : :]) = 0.

It should be noted that a random variable is a numerical quantity whose value is determined by the outcome of a random (chance) experiment. When a random experiment is performed, the totality of outcomes of the experiment forms a set which is called 'sample space' (S) of the experiment.

1. Binomial Distribution.
2. Multinomial Distribution.
3. Negative Binomial Distribution.
4. Poisson Distribution.
5. Hypergeometric Distribution, and
6. Normal Distribution.

Among these the first five distributions are of discrete type and the last one of continuous type. It may also be pointed out that of the six distributions mentioned, the Binomial, Poisson and Normal find much more wider application in practice than the other three. Hence they shall be discussed in detail.

The probabilities corresponding to these results are :

TT	TH	HT	HH
qq	qp	pq	pp
q^2		2qp	p^2

These are the terms of the binomial $(q + p)^2$ because

$(q + p)^2 = q^2 + 2ap + p^2$

In a special case where $p = q = \frac{1}{2}$, we have

$$\left(\frac{1}{2}+\frac{1}{2}\right)^2 + \frac{1}{4}+\frac{1}{2}+\frac{1}{4}.$$

Similarly, if three coins A, B and C are tossed the following are the possible outcomes and the probabilities corresponding to these results are:

ABC	ABC	ABC	ABC	ABC	ABC	ABC	ABC
TTT	TTH	THT	HTT	THH	HTH	HHT	HHH
q^3	q^2p	q^2p	q^2p	qp^2	qp^2	qp^2	p^3

These are the terms of the binomial $(q + p)^3$

$(q + p)^3 = q^3 + 3q^2p + 3qp^2 + p^3$

where $p = q = \frac{1}{2}$, we have $\left[\frac{1}{2}+\frac{1}{2}\right]^3 + \frac{1}{8}+\frac{3}{8}+\frac{3}{8}+\frac{1}{8}.$

These probabilities can be calculated by direct count also. For example, the chance of getting 3 tails in a single toss of 3 coins is 1/8. The chance of getting 2 tails (combined with one head) is 3/8, the chance of getting 1 tall (combined with 2 heads) is 3/8 and the chance of getting no tails is 1/8. In general in n tosses of a coin the probabilities of the various possible events (*i.e.*, obtaining 0, 1, 2, ... n heads) are given by the successive terms of the binomial expansion $(q + p)^n$, which is

$$(q + p)^n = q^n + {}^nC_1q^{n-1}p + {}^nC_2q^{n-2}p^2 + ... + {}^nC_rq^{n-r}p^r + ... p^n.$$

These terms may be listed in the form of a probability distribution table as follows:

Probability Table for Number of Heads

Number of heads X	*Probability* *(r) = P(X = r)*
0	${}^nC_0p^0q^n = q^n$
1	${}^nC_1q^{n-1}p$
2	${}^nC_2q^{n-2}p^2$
3	${}^nC_3q^{n-3}p^3$
r	${}^nC_rq^{n-r}p^r$
⋮	⋮
n	${}^nC_np^nq^0 = p^n$

BINOMIAL DISTRIBUTION

This distribution has been used to describe a wide variety of processes in business and the social sciences as well as other areas. The type of process which gives rise to this distribution is usually referred to as Bernoulli process is series of experimental trials. These assumptions are:

1. An experiment is performed under the same conditions for a fixed number of trials, say, n.
2. In each trial, there are only two possible outcomes of the experiment. For lack of a better nomenclature they are called "success" or "failure". Stated in somewhat different language, the sample space of possible outcomes on each experimental trial is :

$$S = \{\text{failure, success}\}$$

The Binomial Distribution

$$P(r) = {}^nC_r q^{n-r} p^r$$

where p = Probability of success in a single trial

$q = 1 - p$

n = Number of trials

r = Number of successes in n trials

If we want to obtain the probable frequencies of the various outcomes in N sets on n trials, the following expression shall be used:

$$N(q + p)^n$$

$$N(q + p)^n = Nq^n + {}^nC_1\, q^{n-1}\, p + {}^nC_2 q^{n-2}\, p^2 + ... + {}^nC_r q^{n-r}\, p^r + ... + p^n$$

The frequencies obtained by the above expansion are known as expected of theoretical frequencies. On the other hand, the frequencies actually obtained by making experiments are called actual or observed frequencies. Generally, there is some difference between the observed and expected frequencies but the difference becomes smaller and smaller as N increases.

Obtaining Coefficients of the Binomial

For obtaining coefficients from the binomial expansion, the following rules may be remembered. To find the terms of the expansion of $(q + p)^n$

1. The first term is q^n.
2. The second term is ${}^nC_1 q^{n-1}\, p$.
3. In each succeeding term the power of q is reduced by 1 and the power of p is increased by 1.
4. The coefficient of any term is found by multiplying the coefficient of the proceeding term by the power of q in that preceding term, and dividing the products so obtained by one more than the power of p in that preceding term.

When we expand $(q + p)^n$. We get

$$(q + p)^n = q^n + {}^nC_1\, q^{n-1}\, p + {}^nC_2\, q^{n-2}\, p^2 + \ldots\ldots + {}^nC_r\, q^{n-r}\, p^r + \ldots\ldots p^n.$$

where 1. ${}^nC_2$ ..... are called the binomial coefficients. Thus in the expansion of $(q + p)^5$ we will have

$$(q + p)^5 = q^5 + 5q^4p + 10q^3\, p^2 + 10q^2p^3 + 5qp^4 + p^5$$

and the coefficients will be 1, 5, 10, 10, 5, 1.

From the above binomial expansion, the following general relationship should be noted :

(1) The number of terms in a binomial expansion is always n + 1.

(2) The exponents of p and q, for any single term, when added together, always sum to n.

(3) The exponents of q are n, (n – 1), (n – 2)....1, 0, respectively and the exponents of p are 0, 1, 2..(n – 1). n respectively (note : $p^0 = 1$: $q^0 = 1$).

(4) The coefficients for the n + 1 terms of the distribution are always symmetrical ascending to the middle of the series and then descending, when n is odd number, n + 1 is even and the coefficients of the two central terms are identical.

The coefficients of the binomial expansion can be obtained from the Pascal's triangle given below.

Pascal's triangle [showing coefficients of the team $(q + p)^n$]

Row number No. trials n	Binomial Coefficients	No. of possible outcomes 2^n.
1	1 1	2
2	1 2 1	4
3	1 3 3 1	8
4	1 4 6 4 1	16
5	1 5 10 10 5 1	32
6	1 6 15 20 15 6 1	64
7	1 7 21 35 35 21 7 1	128
8	1 8 28 56 70 56 28 8 1	256
9	1 9 36 84 126 126 84 36 9 1	512
10	1 10 45 120 210 252 210 120 45 10 1	1,024

Understanding Pascal's Triangle

1. As the accompanying table lists, the number of each row represents the number of trials of some experiment with two possible types of outcome per trial. Row I represents one trial, such as tossing a coin once. Row two represents two trials such as tossing a coin twice. Row 8 represents, 8 trials such as tossing a coin 8 times.
2. The sum of the actual entries in each row represents the number of possible outcomes. The number is row I sum to 2, there are two possible outcomes of one trial (such as head or tails) if the coin is tossed once the number in row 4 sum to 16, there are 16 possible outcomes of 0 trial.
3. Each individual number in a given row represents the possible number of times some particular outcomes (such as r heads) will occur. The first number in a given row tells how often n particular outcomes (such as n heads) will occur in n trials, the second number tells us how often n – 1 particular outcomes (such as n – 1 heads) will occur in n trials.

Properties of the Binomial Distribution

1. The shape and location of binomial distribution changes as p changes for a given n or as n changes for a given p. As p increases for a fixed n, the binomial distribution shifts to the right.
2. As n increases for a fixed p, the binomial distribution moves to the right, flattens, and spreads out. The mean of the binomial distribution, np obviously increases as n increases with p held constant. For larger n there are more possible outcomes of a binomial experiment and the probability associated with any particular outcome becomes smaller.
3. The mode of the binomial distribution is equal to the value of x which has the largest probability. For example, if $n = 6$ and $p = 0.3$, the mode is equal to 2. While for $n = 6$ and $p = 0.9$ the mode is equal to 6. The mean and mode are equal if np is an integer. For example, when $n = 6$ and $p = 0.50$, the mean and mode are both equal to 3. For fixed n, both mean and mode increase as p increases.
4. If n is large and if neither p nor q is too close to zero, the binomial distribution can be closely approximated by a normal distribution with standardized variable given by $z = \frac{X - np}{\sqrt{npq}}$. The approximation becomes better with increasing n.

Constants of the Binomial Distribution

The mean of the binomial distribution is np and standard deviation $\sqrt{npq}$.

Proof

If p is the probability of success and p the probability of failure in one trial then in n independent trials the probabilities of 0, 1, 2, 3,....n successes are given by the 1st, 2nd 3rd.......n + 1th term of the binomial expansion $(q + p)^n$. Thus we have :

x	p(x)	xp(x)
0	q^n	$0 \times q^n$
1	${}^nC_1\, q^{n\ \ 1}\, p$	$1 \times nq^{n-1}\, p$
2	${}^nC_2\, q^{n\ \ 2}\, p^2$	$\dfrac{2n(n-1)}{2\times 1} q^{n-2}\, p^2$
:	:	:
:	:	:
:	:	:
n	p^n	np^n
	$\Sigma\, p(x) = 1$	

The arithmetic mean $= \dfrac{\Sigma x.p(x)}{\Sigma p(x)}$

$$\Sigma x.p(x) = 0.q^n + nq^{n-1}\, p + \frac{2n(n-1)}{2}\, q^{n-2}\, p^2 + ... np^n$$

$= nq^{n-1}\, p + n\,(n - 1)\, q^{n-2}\, p^2 + ...np^n.$

Taking np common we have

$$= np\left[q^{n-1} + (n-1)\, q^{n-2}\, p\right] + \frac{n(n-1)\,(n-2)}{2\,!} q^{n-3} + ... + p^{n-1}$$

$= np\ (q + p)^{n-1}$

[since the expansion in brackets is the expansion of the binomial $(q + p)^{n-1}$]

$= np\ (1)^{n-1} = np$ $\qquad (\because q + p = 1)$

Thus $\Sigma\, x \,.\, p(x) = np$ ($\because$ the sum probabilities = 1)

Thus the mean of binomial distribution is np.

The standard deviation of binomial distribution is $\sqrt{npq}$.

Proof

σ^2 or $\mu_2 = \nu_2 - \nu^2_1$ (where ν_1 and ν_2 are moments about origin, zero)

$\nu_2 = \Sigma\,(x^2 \,.\, p\,(x))$

$\nu_1 = np$

$$\Sigma\,\{x^2 \,.\, p\,(x)\} = 0^2 \,.\, q^n + 1^2\, nq^{n-1}\, p + \frac{2^2 . n(n-1)}{2\,!}\, q^{n-2}\, p^2$$

$$+ \frac{3^2\, n\,(n-1)\,(n-2)}{3\,!}\, q^{n-2}\, p^3 + ... + n^2\, p^n$$

$$= nq^{n-1} + 2.n\,(n-1)\, q^{n-2} p^2 + \frac{3n(n-1)\,(n-2)}{2 \times 1}\, q^{n-3}\, p^3 + ... + n^2 p^n$$

$$= np\; q^{n-1} + 2\,(n-1)\, q^{n-2} p + \frac{3\,(n-1)\,(n-2)}{2 \times 1}\, q^{n-3}\, p^2 + ... + p^{n-1}$$

Breaking second, third and following terms into pars, we get

$$\Sigma x^2 . p(x) = p\left[\left(q^{n-1} + (n-1) q^{n-2}\, p + \frac{(n-1)(n-2)}{2 \times 1} q^{n-2}\, p^2 + ... + p^{n-1}\right)\right]$$

$$+ \{(n-1)\, q^{n-2} + \frac{(n-1)\,(n-2)}{2 \times 1} q^{n-3}\, p^2 \,....+ (n-1)\, p^{n-1}\}$$

$= np\,(q + p)^{n-1} + (n-1)\,p\,\{q^{n-2} + (n-2)\, q^{n-2}\, p + ... + p^{n-1}\}$

$= np\,[1 + (n-1)\, p\,(q + p)^{n-2}]$

[$\therefore$ $(q + p)^{n-1} = 1$ and the expression to the right is expansion of $(q + p)^{n-2} = 1$]

$= np\,[1 + (n-1)\,p.1]$

$= np\,[1 + np - p]$

$= np + n^2\, p^2 - np^2$

$\mu_2 = \nu_2 - \nu^2_1.$

$= np + n^2p^2 - np^2 - (np)^2$

$= np + n^2p^2 - np^2 - n^2p^2.$

$= np - np^2$

$= np\,(1 - p)$

$= npq \qquad [\because (1 - p) = q)$

$\therefore \sigma \text{ or } \sqrt{\mu_2} = \sqrt{npq}$

Thus the standard deviations of binomial distribution is $\sqrt{npq}$ or the variance is npq.

Similarly,

$\mu_3 = npq\ (q - p)$, and

$\mu_4 = 3n^2\ p^2\ q^2 + npq\ (1 - 6pq)$

From the various moments the value of β_1 and β_2 can also be computed as

$$\beta_1 = \frac{\mu_3^2}{\mu_2^3} = \frac{n^2p^2q^2(q-p)^2}{n^3p^2q^3} = \frac{(q-p)^2}{npq}$$

If β_1 = the distribution is symmetrical, *i.e.*, there is no skewness. Prof. Fisher gave the following measure of skewness :

$$\gamma_1 = \sqrt{\beta_1}$$

If $\gamma_1 = 0$ the distribution is symmetrical, if γ_1 is more than zero the distribution is positively skewed and if it is less than zero the distribution is negatively skewed. It should be noted that γ_1 is a better measure of skewness compared to β_1 because β_1 will always be positive whereas γ_1 can be both positive as well as negative.

$$\beta_2 = \frac{\mu_4}{\mu_2^2} = \frac{3n^2p^2 + npq\,(1-6pq)}{n^2p^2q^2} = 3 + \frac{1-6pq}{npq}$$

Prof. Fisher gave the following measure of kurtosis :

$$\gamma_2 = \beta_2 - 3$$

If the distribution is normal, γ_1 would be zero and γ_2 would also be zero. But the converse is not true, *i.e.*, even if both γ_1 and γ_2 are zero the distribution is not necessarily normal. If γ_2 is positive, the distribution is leptokuritic and if γ_2 is negative the distribution is platykuritic.

The various constants of the binomial distribution can be listed in the following table :

Constants of Binomial Distribution

Mean = np

Standard Deviation = $\sqrt{npq}$

First Moment or $\mu_1 = 0$

Second Moment or μ_2 = npq

Third Moment or μ_3 = npq (q − p)

Fourth Moment or $\mu_4 = 3n^2p^2q^2 + npq\,(1 - 6pq)$

$$\beta_1 = \frac{(q-p)^2}{npq} \qquad \beta_2 = 3 + \frac{1 - 6\,pq}{npq}$$

Importance of the Binomial Distribution

1. The outcome or results of each trial in the process are characterised as one of two types of possible outcomes. In other words, they are attributes.
2. The possibility of outcome of any trial does not change and is independent of the results of previous trials.

The following examples will illustrate the applications of binomial distribution.

Example 1:

The incidence of a certain disease is such that on the average 20% of workers suffer from it. If 10 workers are selected at random, find the probability that

(i) Exactly 2 workers suffer from the disease,

(ii) not more than 2 workers suffer from the disease.

Calculate the probability upto fourth decimal place.

Solution:

Probability that a worker suffers from a disease $= \frac{20}{100} = \frac{1}{5}$ *i.e.* $p = \frac{1}{5}$ and $q = \frac{4}{5}$.

By binomial probability law, the probability that out of 10 workers, x workers suffer from a disease is given by :

$$P(r) = {}^nC_r\, q^{n-r}\, p^r = 10C_r \left(\frac{4}{5}\right)^{10-r} \left(\frac{1}{5}\right)^r$$

$$= 10C_r \frac{4^{10-r}}{5^{10}};\ r = 0, 1, 2, ...10.$$

(i) The required probability that exactly 2 workers will suffer from the disease is given by :

$$P_{(2)} = 10C_r \frac{4^{10-2}}{5^{10}} = \frac{45 \times 4^8}{5^{10}} = 0.302$$

(ii) The required probability that not more than 2 workers will suffer from the disease is given by :

$$P(0) + P(1) + P(2) = \frac{1}{5^{10}}[10\ C_0\ 4^{10} + 10C_1\ 4^9 + 10C_2\ 4^8] = 0.678$$

Example 2:

If the probability of defective bolts is 0.1, find (a) the mean and standard deviation for the distribution of defective bolts in a total of 500, and (b) the moment coefficient of skewness and kurtosis of the distribution.

Solution:

(a) $p = 0.1, \quad n = 500$

Mean = np = 500 × .1 = 50.

Thus we can expect 50 bolts to be defective.

$$\sigma = \sqrt{npq}$$

$n = 500$, $p = 0.1$ and $q = 0.9$

$$\sigma = \sqrt{500 \times 0.1 \times 0.9} = 6.71$$

(b) Moment coefficient of skewness, *i.e.* γ_1.

$$\gamma_1 = \sqrt{\beta_1} = \sqrt{\frac{(q-p)^2}{npq}}$$

$$= \frac{q-p}{\sqrt{npq}} = \frac{(0.9 - 0.1)}{6.71} = \frac{0.8}{6.71} = 0.119.$$

Since γ_1 is more than zero the distribution is positively skewed. However, skewness is very moderate.

Moment coefficient of kurtosis

$$\gamma_2 = \beta_2 - 3$$

$$\beta_2 = 3 + \frac{1 - 6pq}{npq} = 3 + \frac{1 - 6(0.1)(0.9)}{44.9}$$

$$= 3 + \frac{0.46}{44.9} = 3.01,$$

$$\gamma_1 = 3.01 - 3 = +0.01$$

Since γ_2 is positive the distribution is platykurtic

$$(q + p)^5 = q^5 + 6q^5p + 15q^4 p^2 + 20q^2p^2 + 15q^2 p^4 + 6pq^5 + p^6.$$

Fitting a Binomial Distribution

When a binomial distribution is to be fitted to observe data, the following procedure is adopted :

(1) Determine the values of p and q. If one of these values is known the other can be found out by the simple relationship p = (1 – q), and q = (1 – p). When p and p are equal the distribution is symmetrical, for p and q may be interchanged without alternating the value of any terms, and consequently terms equidistant from the two ends of the series are equal. If p and q are unequal, the distribution is skew. If p is less than 1/2, the distribution is positively skewed and when p is more than 1/2 the distribution is negatively skewed.

(2) Expand the binomial $(q + p)^n$. The power n is equal to one less than the number of terms in the expanded binomial. Thus when two coins are tossed (n = 2) there will be three terms in the binomial, Similarly, when four coins are tossed (n = 4) there will be five terms, and so on.

(3) Multiply each term of the expanded binomial by N (the total frequency), in order to obtain the expected frequency in each category.

The following examples shall illustrate the procedure :

Example 3:

The screws product by a certain machine were checked by examining samples of 12. The following table shows the distribution of 128 samples according to the number of defective items they contained :

No. of defectives in a sample of 12 :	*0*	*1*	*2*	*3*	*4*	*5*	*6*	*7*	*Total*
No. of samples:	*7*	*6*	*19*	*35*	*30*	*23*	*7*	*1*	*128*

Fit a binomial distribution and find the expected frequencies if the chance of machine being defective is 1/2. Find the mean and variance of the fitted distribution.

Solution:

On the hypothesis that the coins are unbiased, N = 128 and n = 7. The probability of 0, 1, 2........7 defectives will be given by the expansion $\left(\frac{1}{2}+\frac{1}{2}\right)^7$

$$\left(\frac{1}{2}+\frac{1}{2}\right)^7 = {}^7C_1\left(\frac{1}{2}\right)^4\left(\frac{1}{2}\right) + {}^7C_2 + \left(\frac{1}{2}\right)^5\left(\frac{1}{2}\right)^2 + {}^7C_3\left(\frac{1}{2}\right)^4\left(\frac{1}{2}\right)^5$$

$$+{}^7C_2\left(\frac{1}{2}\right)^3\left(\frac{1}{2}\right)^4 {}^7C_5\left(\frac{1}{2}\right)^2\left(\frac{1}{2}\right)^5+{}^7C_6\left(\frac{1}{2}\right)\left(\frac{1}{2}\right)^6+{}^7C_r\left(\frac{1}{2}\right)^0\left(\frac{1}{2}\right)^7$$

$$=\left(\frac{1}{2}\right)^7 = [1 + 7 + 21 + 35 + 35 + 21 + 7 + 1]$$

In order to obtain frequencies, we will have to multiply each term by N, *i.e.* 128. Hence :

$$128\left(\frac{1}{2}+\frac{1}{2}\right)^7 = 128 \times \frac{1}{128}\ (1 + 7 + 21 + 35 + 35 + 21 + 7 + 1)$$

Thus the expected frequencies are :

X	0	1	2	3	4	5	6	7
fe	1	7	21	35	35	21	7	1

Mean and variance of the fitted distribution :

Mean of he binomial distribution is np and standard deviation $\sqrt{npq}$

Hence mean = $7 \times \frac{1}{2} = 3.5$ and variance = $\frac{1}{2} \times \frac{1}{2} \times 7 = 1.75$.

Example 4:

Twelve dice were thrown 4,096 times. Each 4, 5 or 6 spot appearing was considered to be a success while a 1, 2 or 3 spot was a failure. Calculate the theoretical frequencies for 0, 1, 2...., 12 successes.

Solution:

There are 4,096 trials. Since either 4, 5 or 6 is considered a success we have q = p = 1/2.

The terms of the binomial $(q + p)^n$ will give the probabilities of 0, 1, 2,.....n successes.

Here n = 12, q = 1/2 and p = 1/2.

$\therefore$ By expanding $4096\left(\frac{1}{2}+\frac{1}{2}\right)^{12}$

We get frequencies corresponding to 0, 1, 2,........, 12 successes.

$$4096\left(\frac{1}{4096}+\frac{12}{4096}+\frac{66}{4096}+\frac{220}{4096}+\frac{495}{4096}+\frac{792}{4096}+\frac{924}{4096}\right.$$

$$\left.+\frac{792}{4096}+\frac{495}{4096}+\frac{220}{4096}+\frac{66}{4096}+\frac{12}{4096}+\frac{1}{4096}\right)$$

f_0 will denote observed frequencies and f_e expected frequencies. The observed frequencies cannot be in fractions but the expected frequencies may be in fractions. However, they may be approximated to the whole number.

The result can be tabulated as follows :

No. of Successes	Theoretical frequencies	No. of Successes	Theoretical frequencies
0	1	7	792
1	12	8	495
2	66	9	220
3	220	10	66
4	495	11	12
5	792	12	1
6	924		

Example 5:

A student obtained the following answer to a certain problem given to him. Mean = 2.4; Variance = 3.2 for a binomial distribution. Comment on the result.

Solution:

The mean of binomial distribution is np and variance npq. We are given mean = np = 2.4

Variance = npq   2.4q = 3.2 or

$$q = \frac{3.2}{2.4} = 1.333$$

Since the value of q is greater than 1, the given results are inconsistent.

Example 6:

A sample of 3 lines is selected at random from a box containing 12 items of which 3 are defective. Find the possible number of defective combinations of the said 3 selected items along with probability of a defective combination.

Solution:

The usual notions we have :

$$n = 3, p = \frac{3}{12} = \frac{1}{4}, q = 1 - \frac{1}{4} = \frac{3}{4}$$

Then by the binomial probability law, the probability that there are r defective items is a given by:

No. of defective items	Probability
0	$3C_0\left(\frac{1}{4}\right)^0\left(\frac{3}{4}\right)^3 = \frac{27}{64}$
1	$3C_1\left(\frac{1}{4}\right)^1\left(\frac{3}{4}\right)^2 = \frac{27}{64}$
2	$3C_2\left(\frac{1}{4}\right)^2\left(\frac{3}{4}\right)^1 = \frac{9}{64}$
3	$3C_3\left(\frac{1}{4}\right)^3\left(\frac{3}{4}\right)^0 = \frac{1}{64}$

Example 7:

The following data show the number of seeds germinating out of 10 on damp filter for 80 set of seeds. Fit a binomial distribution to this data :

X:	*0*	*1*	*2*	*3*	*4*	*5*	*6*	*7*	*8*	*9*	*10*
Y:	*6*	*20*	*28*	*12*	*8*	*6*	*0*	*0*	*0*	*0*	*0*

Solution:

Fitting Binomial Distribution

X	f	fX
0	6	0
1	20	20
2	28	56
3	12	36
4	8	32
5	6	30
6	0	0
7	0	0
8	0	0
9	0	0
10	0	0
	N = 80	Σ fX = 174

$$\overline{X} = \frac{174}{80} = 2.175$$

But mean $= np = \frac{174}{80}$ $\therefore p = \frac{174}{800} = 0.2175$

$\therefore q = 1 - p = 0.7825.$

Hence the binomial distribution to be fitted the data is

$80\ (0.7825 + 0.2175)^{10}$

The theoretical frequencies are the successive terms in the expansion of $80\ (0.7825 + 0.2175)^{10}$ and are tabulated here.

X	Theoretical frequencies $N \times {}^nC_r\, q^{n-r}\, p^r$.
0	$80 \times (.7825)^{10} = 6.9$
1	$80 \times 10\ (.7825)^9\ (.2175)^1 = 19.1$
2	$80 \times 45\ (.7825)^8\ (.2175)^2 = 24.0$
3	$80 \times 120\ (.7825)^7\ (.2175)^3 = 17.8$
4	$80 \times 210\ (.7825)^6\ (.2175)^4 = 8.6$
5	$80 \times 252\ (.7825)^5\ (.2175)^5 = 2.9$
6	$80 \times 210\ (.7825)^4\ (.2175)^6 = 0.7$
7	$80 \times 120\ (.7825)^3\ (.2175)^7 = 0.1$
8	$80 \times 45\ (.7825)^2\ (.2175)^8 = 0.0$
9	$80 \times 10\ (.7825)^1\ (.2175)^9 = 0.0$
10	$80 \times (.2175)^{10} = 0.0$
	Total = 80.1

Example 8:

Eight coins are tossed at a time 256 times. Number of heads observed at each throw is recorded and the results are given below. Find the expected frequencies. What are the theoretical values of mean and standard deviation? Calculate also the mean and S.D. of the observed frequencies.

No. of heads at a throw	***Frequency***	***No. of heads at a throw***	***Frequency***
0	*2*	*5*	*56*
1	*6*	*6*	*32*
2	*30*	*7*	*10*
3	*52*	*8*	*1*
4	*67*		

Solution:

The chance of getting a head in a single throw of one coin is $\frac{1}{2}$

Hence $p = \frac{1}{2}$. $q = \frac{1}{2}$, n = 8, N = 256.

By expanding $256\left(\frac{1}{2}+\frac{1}{2}\right)^8$ we shall get the expected frequencies of 1, 2,......, 8 heads (successes)

No. of head (X)	Frequency = N × $^nC_r\, q^{n-r}\, p^r$.
0	$256\left(\frac{1}{2}\right)^8 = 1$
1	$256 \times {}^8C_1\left(\frac{1}{2}\right)^1\left(\frac{1}{2}\right)^7 = 8$
2	$256 \times {}^8C_2\left(\frac{1}{2}\right)^2\left(\frac{1}{2}\right)^6 = 28$
3	$256 \times {}^8C_3\left(\frac{1}{2}\right)^3\left(\frac{1}{2}\right)^5 = 56$
4	$256 \times {}^8C_4\left(\frac{1}{2}\right)^4\left(\frac{1}{2}\right)^{64} = 70$
5	$256 \times {}^8C_5\left(\frac{1}{2}\right)^5\left(\frac{1}{2}\right)^3 = 56$
6	$256 \times {}^8C_6\left(\frac{1}{2}\right)^6\left(\frac{1}{2}\right)^2 = 28$
7	$256 \times {}^8C_7\left(\frac{1}{2}\right)^7\left(\frac{1}{2}\right)^1 = 8$
8	$256 \times \left(\frac{1}{2}\right)^8 = 1$
	Total = 256

The mean of the above distribution is np $= 8 \times \frac{1}{2} = 4$.

The standard deviation is $\sqrt{npq} = \sqrt{\frac{1}{2} \times \frac{1}{2} \times 8} = \sqrt{2} = 1.414$

These are the mean and standard deviation of the expected frequency distribution. The mean and standard deviation of the observed frequency distribution shall be :

X	f	d	fd	fd²
0	2	–4	–8	32
1	6	–3	–18	54
2	30	–2	–60	120
3	52	–1	–52	52
4	67	0	0	0
5	56	+1	+56	56
6	32	+2	+64	128
7	10	+3	+30	90
8	1	+4	+4	16
	N = 256		Σ fd = 16	Σ fd² = 548

$$\overline{X} = A + \frac{\Sigma fd}{N}$$

$$= 4 + \frac{16}{256} = 4.0625$$

$$\sigma = \sqrt{\frac{\Sigma fd^2}{N} - \left(\frac{\Sigma fd}{N}\right)^2}$$

$$= \sqrt{\frac{548}{256} - \left(\frac{16}{256}\right)^2}$$

$$= \sqrt{2.141 - 0.004}$$

$$= \sqrt{2.137} = 1.462.$$

The Multinomial Distribution

An important generalization of the binomial distribution arises where there are more than two possible outcomes for each trial, the probabilities of the various outcomes remain in same for each trial, and the trials are independent. Whereas in case of binomial distribution, there are only two possible outcomes on each experimental trial, in the multinomial distribution there are more than two possible outcomes on each trial. For example, when a symmetric six-side dice is rolled there are six possible outcomes. 1, 2, 3, 4, 5 or 6. The assumptions underlying the multinomial distribution are analogous to the binomial distribution. These are :

(1) An experiment is performed under the same conditions for a fixed number of trials, say, n.

(2) There are K mutually exclusive and exhaustive outcomes of the experiment which may be referred to E_1, E_2, E_3.... E_k. Thus the sample space of possible outcomes on each trial shall be :

$$S = \{E_1, E_2, E_3, ... E_k\}$$

(3) The respective probabilities of the various outcomes, *i.e.*, E_1, E_2, E_3 .....E_k denoted by p_1, p_2, p_3, ... p_k respectively remain constant from trial to trial.

$$p_1 + p_2 + p_3 + ...p_k = 1$$

(4) The trials are independent.

Under the above assumptions, the probability that there will be r_1 occurrence of E_1, r_2 occurrence of E_2, ... r_k occurrence of E_k in n trials is given by :

$$P(r_1, r_2, ...r_k) = \frac{n!}{r_1!\, r_2!\, r_k!} p_1^{r1} p_2^{r2} .....p_k^{r3}$$

Where $r_1 + r_2 + ... + r_k = n$ and $p_1 = p_2 + ... p_k = 1$.

The general form of the multinomial distribution is

$$(p_1 + p_2 + ...p_k)^n$$

Example 1:

A dice is role five times. Find the probability of getting a1, a2 and three other numbers.

Solution:

The multinomial mode is appropriate for this problem. Applying the formula :

$$P(r_1, r_2, ...r_k) = \frac{n!}{r_1!\, r_2!\, r_k!} p_1^{r1} p_2^{r2} .....p_k^{kr}$$

$r_1 = 1, r_2 = 1, r_3 = 3$. Hence, the probability of obtaining a1, a2 and three other number is :

$$p = \frac{5!}{1!\,1!\,3!}\left(\frac{1}{6}\right)\left(\frac{1}{6}\right)\left(\frac{4}{6}\right)^2 = \frac{40}{243} = 0.165.$$

Example 2:

A community consists of 50 percent Hindus, 30 percent Muslims and 20 percent Sikhs. If a sample of six individuals is selected at random, what is the probability that two are Hindus, three are Muslims and one is a Sikh?

Solution:

The answer is provided by the multinomial distribution. The probabilities of persons of these races are respectively 0.50, 0.30 and 0.20. Hence the required probability shall be given by :

$$p = \frac{6!}{2!\,3!\,1!}(0.5)^2(0.3)^2(0.2) = 0.081.$$

Negative Binomial Distribution

The negative binomial distribution is very much similar to the binomial probability model. It is applicable when the following conditions hold good:

1. An experiment is performed under the same conditions till a fixed number of successes, say C_2 is achieved.
2. The result of each experiment can be classified into one of the two categories, success or failure.
3. The probability p of success is the same for each experiment.
4. Each experiment is independent of all the others.

Consider a sequence of Bernoulli trials with p as the probability of success. In the sequences success and failure will occur randomly and in each trial the probability of success will be p and that of a failure will be–p=q. Let us investigate how much time will be taken to reach the rth success. Here r is fixed, let the number of failures preceding the rth success be x (= 0, 1, 2,...). Then the total number of trials to be performed to reach the rth success will be x + r. (Here x is a chance variable.) Then the probability that rth success occurs at (x + r)th trial equals the probability that exactly x failures (r + x – 1) trials there are (r – 1) successes and the (r + x)th trial is a success; the corresponding probabilities are $\binom{hr + x - 1}{r - 1} p^{r-1} 1^{x}$ and p respectively. The probability of the compound event will be :

$$\binom{r+x-1}{r-1} p^{r-1} q^{x} . p = \binom{r+x-1}{r-1} p^{r} q^{x} \qquad ...(i)$$

Since the events are independent.

The probabilities are generally denoted by f (x, r, p). The probabilities given by (i) are known as 'negative binomial probabilities' and the law is known as the negative binomial probability law.

Definition : *A discrete random variable X is said to follow 'negative binomial distribution' if it takes only non-negative values and its probability mass function is given by*

$$f(x,r,p) = \binom{r+x-1}{r-1} p^{r} q^{x};\ x = 0,1,2,.....$$

$$= 0 \text{ ; otherwise} \qquad ...(ii)$$

The fixed quantities r and p are the two parameters of the distribution.

It can be easily provided that Σ f (x, r, p) summation extends over all the possible values of x, is equal to unity therefore (2) is a probability distribution.

By simple calculations, it can be shown that

$$\binom{r+x-1}{r-1} = (-1)^{x} \binom{-r}{x}$$

Then (2) can be rewritten as

$$f(x,r,p) = \binom{-r}{x} p^{r} (-q)^{x},\ x = 0,1,2,....$$

Which is the (x + 1)th term in the expansion of $p^r (1 - q)^{-r}$ and it is the binomial theorem with negative index. Therefore the terms given by (2) are the successive terms of the expansion of $p^r (1 - q)^{-r}$, hence the name 'Negative Binomial Distribution'.

Example:

Suppose that 30% of the items taken from the end of a production line are defective. If the items taken from the line are checked until 6 defective items are found, what is the probability that 12 items are examined ?

Solution:

Suppose the occurrence of a defective item is a success. Then our question reduces to find the probability that there will be (12 – 6) = 6 failures preceding the 6th success, the probability of success in a trial being 0.30. Then by negative binomial probability law, the required probability will be given by :

$$\binom{6+6-1}{6-1}(.30)^6(.70)^6 = 0.396$$

POISSON DISTRIBUTION

Poisson distribution is a discrete probability distribution and is very widely used in statistical work. It was developed by a French mathematician. *Simeon Denis Poisson* (1781-1840), in 1837. Poisson distribution may be expected in cases where the chance of any individual event being a success is small. The distribution is used to describe the behaviour of rare events such as the number of accidents on road, number of printing mistakes in a book, etc., and has been called "the law of improbable events". In recent years the statisticians have had a renewed interest in the occurrence of comparatively rare events, such as serious floods, accidental release of radiation from a nuclear rector, and the like.

The Poisson distribution is defined as :

The Poisson Distribution

$$P(r) = \frac{e^{-m} m^r}{r!}$$

where r = 0, 1, 2, 3, 4,.......

e = 2.7183 (the base of natural Logarithms)

m = the mean of the Poisson distribution, *i.e.*, np or the average number of occurrences of an event.

The Poisson distribution is a discrete distribution with a single parameter m. As m increases, the distribution shifts to the right. This is illustrated in the following diagram for 4 values from m = 0.3 to m = 4.0.

All Poisson probability distributions are skewed to the right. This is the reason why the Poisson probability distribution has been called the probability distribution of rare events (the probabilities tend to be high for small numbers of occurrences).

The Poisson probability Distribution is concerned with certain processes that can be described by a discrete random variable. The probabilities of 0. 1, 2,.... successes are given by the successive terms of the expansion.

= m.

Thus σ^2 or $\mu_2 = m$ and $\sigma = \sqrt{m}$.

In a similar manner we can show that for Poisson distribution $\mu_2 = m$ and $\mu_4 = m + 3m^2$.

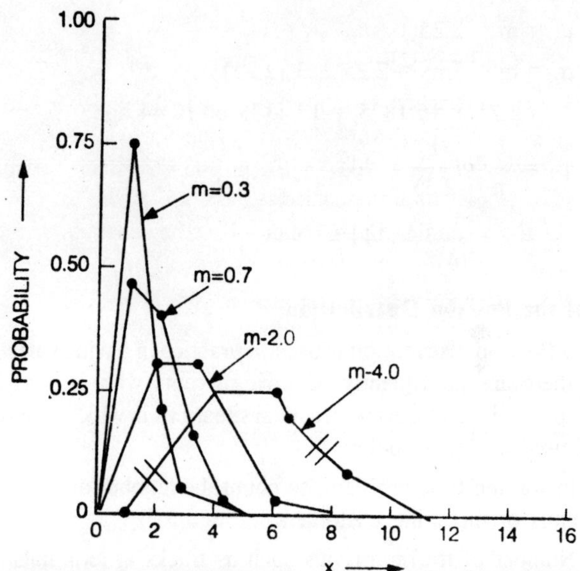

Constants of Poisson Distribution

$$\mu_1 = 0,\ \mu_2 = m,\ \mu_3 = m,\ \mu_4 = m + 3m^2$$

$$\beta_1 = \frac{\mu_3^2}{\mu_2^3} = \frac{m^2}{m^3} = \frac{1}{m}$$

$$\beta_2 = \frac{\mu_4}{\mu_2^2} = \frac{m + 3m^2}{m^2} = 3 + \frac{1}{m}$$

One great advantage of the Poisson distribution is that we need only the value of mean in order to compute the values of various constants. This shall be clear from the following illustration.

Example 1:

The mean of the Poisson distribution is 2.25. Find the other constants of the distribution.

Solution:

We are given mean or m = 2.25

$$\sigma = \sqrt{m} = \sqrt{2.25} = 1.5$$

$$\mu_1 = 0$$
$$\mu_2 = m = 2.25$$
$$\mu_3 = m = 2.25$$
$$\mu_4 = m + 3m^2 = 2.25 + 3\ (2.25)^2$$
$$= 2.25 + 15.1875 = 17.4375 \text{ or } 17.44 \text{ app.}$$
$$\beta_1 = \frac{1}{m} = \frac{1}{2.25} = .444.$$
$$\beta_2 = 3 + \frac{1}{m}\ 3 + .444 = 3.444.$$

Role of the Poisson Distribution

The Poisson distribution is used in practice in a wide variety of problems where there are infrequently occurring events with respect to time, area, volume or similar units. Some practical situations in which Poisson distribution can be used are given below :

(1) In waiting-time problems to count the number of incoming telephone calls or incoming customers.

(2) Number of traffic arrivals such as trucks at terminals, aeroplanes at airports, ships at docks and so forth.

(3) It is used in quality control statistics to count the number of defects of an item.

(4) In biology to count the number of bacteria.

(5) In physics to count the number of particles emitted from a radioactive substance.

(6) In insurance problems to count the number of casualties.

(7) To model the distribution of the number of persons joining a queue (a line) to receive a service or purchase of a product.

(8) In problems dealing with the inspection of manufactured products with the probability that any one piece is defective is very small and the lots are very large, and

(9) The number of typographical errors per page in typed material, number of deaths as a result of road accidents, etc.,

(10) In determining the number of deaths in a district in a given period, say, a year, by a rare disease.

In general, the Poisson distribution explains the behaviour of those discrete variates where the probability of occurrence of the event is small and the total number of possible cases is sufficiently large.

Example 2:

***(a)** Suppose on an average 1 house in 1,000 in a certain district has a fire during a year. If there are 2,000 houses in that district, what is the probability that exactly 5 houses will have a fire during the year?*

Solution:

Applying the Poisson distribution

$$\overline{X} = np$$

$$n = 2000, p = \frac{1}{1000}$$

$$n = 2000 \times \frac{1}{1000} = 2$$

$$P(r) = e^{-m}\frac{m^r}{r!}, r = 0,1,2,......$$

Here $m = 2, r = 5$ and $e = 2.7183$

$$\therefore \quad P(5) = \frac{2.71823^{-2} \times 2^5}{5!}$$

$$= \frac{\text{Rec.}\left[\text{anti log}\left(2 \times \log 2.78183\right)\right] \times 32}{5 \times 4 \times 3 \times 2 \times 1}$$

$$= \frac{\text{Rec.}\left[\text{anti log}\left(2 \times .4343\right)\right] \times 32}{120}$$

$$= \frac{\text{Rec.}\left[\text{anti log}\left(.8686\right)\right] \times 32}{120}$$

$$= \frac{\text{Rec.}\left[7.389\right] \times 32}{120} = \frac{.1352 \times 32}{120} = .036.$$

***(b)** Ten percent of the tools produced in a certain manufacturing process turn out to be defective. Find the probability that in a sample of 10 tools chosen at random, exactly two will be defective by using (a) the binomial distribution, (b) the Poisson approximation to the binomial distribution.*

Solution:

Probability of a defective tool or $p = 0.1$

(a) When a binomial distribution is used probability of 2 defectives in 10 is given by

$$^{10}C_2\ (.1)^2\ (.9)^8 = 0.1937$$

(b) When Poisson distribution is used probability of 2 defectives is given by

$$P(2) = \frac{e^{-m}\, m^2}{2\,!}$$

where m = np = 10 (0.1) = 1.

$$\frac{e^{-1}}{2\,!} = \frac{e^{-1}}{2} = \frac{1}{2e} = 0.184.$$

In general the approximation is good if

$$p \le 0.1 \text{ and } m = np < 5.$$

$$e^{-m}\left(1 + m + \frac{m^2}{2\,!} + \frac{m^3}{3\,!} \ldots\ldots + \frac{m^r}{r\,!} \ldots\ldots + \right)$$

This can be written in a tabular form as follows :

No. of successes (x)	Probability p(x)	No. of successes (x)	Probability p(x)
0	e^{-m}	4	$\frac{m^4 e^{-m}}{4\,!}$
1	me^{-m}	:	:
2	$\frac{m^2 e^{-m}}{2\,!}$	r	$\frac{m^r e^{-m}}{r\,!}$
3	$\frac{m^3 e^{-m}}{3\,!}$	:	:

The above table gives probabilities. If we want to know the expected number of occurrences for different successes. We have to multiply each term by N_2. *i.e.*, the total number of observations.

Constants of the Poisson Distribution

Since p is very small in case of Poisson distribution, the value of q is almost equal to 1. The constants of the Poisson distribution can thus be easily obtained by putting 1 in place of q in the constants of the binomial distribution.

The various constants of the Poisson distribution are :

The mean of the Poisson distribution = m

The standard deviations is $\sqrt{m}$ or μ_2 = m.

Proof:

The Poisson distribution is given as :

No. of successes	*(x)*	0	1	2	3	4.....
Probability	*p(x)*	e^{-m}	$\frac{me^{-m}}{1!}$	$\frac{m^2 e^{-m}}{2!}$	$\frac{m^3 e^{-m}}{3!}$	$\frac{m^4 e^{-m}}{4!}$ ....

Find the mean and variance.

Calculation of Mean and Variance

x	p(x)	x_1 p(x)
0	e^{-m}	0
1	me^{-m}	me^{-m}
2	$\frac{m^2 e^{-m}}{2!}$	$\frac{m^2 e^{-m}}{2} \times 2$
3	$\frac{m^3 e^{-m}}{3!}$	$\frac{m^3 e^{-m}}{3 \times 2 \times 1} \times 3$
4	$\frac{m^4 e^{-m}}{4!}$	$\frac{m^4 e^{-m}}{4 \times 3 \times 2} \times 4$

Mean = Σ x . p(x)

$$\Sigma\, xp = 0 + me^{-m} + m^2e^{-m} + \frac{m^3e^{-m}}{2 \times 1} + \frac{m^4 e^{-m}}{3 \times 2} + ...$$

$$= me^{-m}\left(1 + m + \frac{m^2}{2!} + \frac{m^3}{3!} + ...\right)$$

$$= me^{-m}.\ e^{m} = m \quad \left(\because e^{-m} = 1 + m + \frac{m^2}{2!} + \frac{m^3}{3!} + ...\right)$$

Hence the mean of Poisson distribution is m.

Variance or $\mu_2 = \nu_2 - \nu^2_1$.

[where ν_1 and ν_2 denote moment about origin zero]

$\nu_2 = \Sigma\ \{x^2 . p(x)\}$

x	p(x)	$x^2 \cdot p(x)$
0	e^{-m}	0
1	me^{-m}	me^{-m}
2	$\frac{m^2 e^{-m}}{2!}$	$\frac{m^2 e^{-m}}{2} \times 4$
3	$\frac{m^3 e^{-m}}{3!}$	$\frac{m^3 e^{-m}}{3 \times 2 \times 1} \times 9$
4	$\frac{m^4 e^{-m}}{4!}$	$\frac{m^4 e^{-m}}{4 \times 3 \times 2} \times 16$
:	:	:
:	:	:
r		$\frac{m^r e^{-m}}{r!}$

$$\Sigma\, x^2\, p(x) = 0 + me^{-m} + \frac{2m^2e^{-m}}{1!} + 3\,\frac{m^3e^{-m}}{2!} + 4\,\frac{m^4e^{-m}}{3!} + \ldots.$$

$$= m^{e-m}\left(1 + 2m + 3\frac{m^2}{2!} + 4\,\frac{m^3}{3!} + \ldots.\right)$$

Breaking each term within brackets into two parts each we have

$$\Sigma\, x^2\, p(x) = me^{-m}\left(1 + m + \frac{m^2}{2!} + \frac{m^3}{3!} + \ldots.\right)$$

$$+\left(m + 2\,\frac{m^2}{2!} + 3\,\frac{m^3}{3!} + \ldots.\right)\Bigg\}$$

$$= me^{-m}\left[e^m + m\left(1 + m + \frac{m^2}{2!} + \ldots.\right)\right]$$

$$= me^{-m}\,(e^m + me^m)$$

$$= me^{-m}\,.\,e^m\,(1 + m)$$

$$= me^2\,(1 + m)$$

$$= m + m^2$$

$$\mu_2 = \nu_2 - \nu_1^2$$

$= m + m^2 - (m)^2 \qquad [\because v_1 = + m)$

$$\overline{X} = \frac{\Sigma f X}{N} = \frac{145}{330} = 0.439$$

$P_{(0)} = e^{-m} = e^{-0.439} = 0.6447$ (from the table)

$N(P_0) = P_{(0)} \times N = .6447 \times 330 = 212.75$

$N(P_1) = N(P_0) \times m = 212.75 \times .439 = 93.4$

$$N(P_2) = N(P_1) \times \frac{m}{2} = 93.4 \times \frac{.439}{2} = 20.5$$

$$N(P_3) = N(P_2) \times \frac{m}{3} = 20.5 \times \frac{.439}{3} = 3.0$$

$$N(P_4) = N(P) \times \frac{m}{4} = 3.0 \times \frac{.439}{4} = 0.33$$

Thus the expected frequencies as per Poisson distribution are:

No. of defects	0	1	2	3	4
No. of units	212.75	93.40	20.50	3.00	0.33

(b) *The following table the number of days in a 50 day period during which automobile accidents occurred in a certain part of a city. Fit a Poisson distribution to the data.*

No. of defects	*0*	*1*	*2*	*3*	*4*
No. of units	*19*	*18*	*8*	*4*	*1*

Solution:

Fitting of Poisson Distribution

X	f	fX
0	19	0
1	18	18
2	8	16
3	4	12
4	1	4
	N = 50	Σ fX = 50

$$m = \frac{\Sigma f X}{N} = \frac{50}{50} = 1$$

$P_{(0)} = e^{-m} = 2.7183^{-1} = 0.36788$ (from the table)

$N(P_0) = P_{(0)} \times N = .36788 \times 50 = 18.394$ or 18.4

$N(P_1) = N(P_0) \times m = 18.394 \times 1 = 18.394$ or 18.4

$$N(P_2) = N(P_1) \times \frac{m}{2} = \frac{18.394}{2} = 9.197 \text{ or } 9.2$$

$$N(P_3) = N(P_2) \times \frac{m}{3} = \frac{9.197}{3} = 3.066 \text{ or } 3.1$$

$$N(P_4) = N(P_3) \times \frac{m}{4} = \frac{3.066}{4} = 0.766 \text{ or } 0.8$$

Example 3:

The following mistakes per page were observed in a book:

No. of defects	*0*	*1*	*2*	*3*	*4*
No. of units	*211*	*90*	*19*	*5*	*0*

Solution:

Fitting of Poisson Distribution

X	f	fX
0	211	0
1	90	90
2	19	38
3	5	15
4	0	0
	N = 50	Σ fX = 50

$$\overline{X} = \frac{\Sigma f X}{N} = \frac{143}{325} = 0.44$$

Mean of the distribution or $m = .44$

$P_{(0)} = e^{-m} = 2.7183^{-.44}$

= Rec. [antilog (.44 log 2.7183)] = Rec. [antilog (.44 × 4343)]

= Rec. [antilog 1.1910] = Rec. 1.552 = 6443

$N(P_0) = .6443 \times 325 = 209.40$

$$N(P_1) = .6443 \times \frac{m}{1} = 209.4 \times .44 = 92.14$$

$$N(P_2) = N(P_1) \times \frac{m}{2} = 92.14 = \frac{.44}{2} = 92.14 \times .22 = 20.27$$

$$N(P_3) = N(P_1) \times \frac{m}{2} = 92.14 = \frac{.44}{2} = 92.14 \times .22 = 20.27$$

$$N(P_4) = N(P_3) \times \frac{m}{4} = 2.97 \times \frac{.44}{4} = 2.97 \times .11 = 0.33$$

The expected frequencies of Poisson distribution are:

X	0	1	2	3	4	
f	209.40	92.14	20.27	2.97	0.33	= 325.14

Note: A rough check on the accuracy of result is that the total of the expected frequencies should be equal to the total of the observed frequencies. For example, in the above case that total.

Fitting a Poisson Distribution

The process of fitting a Poisson distribution is very simple. We have just to obtain the value of m. *i.e.* the average occurrence and calculate the frequency of 0 success. The other frequencies can be very easily calculated as follows:

$$N(P_0) = Ne^{-m}$$

$$N(P_1) = N(P_0) \times \frac{m}{1}$$

$$N(P_2) = N(P_1) \times \frac{m}{2}$$

$$N(P_3) = N(P_3) \times \frac{m}{3} \text{ etc.}$$

A goodness-of-fit' test will confirm whether or not the fit is close enough to justify the belief that the distribution is of the Poisson type.

Example 4:

The number of defects per unit in a sample of 330 units of manufactured product was found as follows:

No. of defects	*0*	*1*	*2*	*3*	*4*
No. of units	*214*	*92*	*20*	*3*	*1*

Fit a Poisson distribution to the data and test for goodness of fit. (Given $e^{-0.439} = 0.6447$*)*

Solution:

Fitting of Poisson Distribution

X	f	fX
0	214	0
1	92	92
2	20	40
3	3	9
4	1	4
	N = 300	Σ fX = 145

of expected frequencies is 325.14 and the observed total is 325. The slight difference is due to approximation.

(b) *In a certain factory manufacturing razor blades, there is small chance 1/50 for any blade to be defective. The blades are placed in packets, each containing 10 blades. Using the Poisson distribution, calculate the approximate number of packets containing not more than 2 defective blades in a consignment of 10,000 packets.*

Solution:

$$N = 10{,}000, \; p = \frac{1}{50}, n = 10$$

$$m = np = 10 \times \frac{1}{50} = 0.2$$

$$P_{(0)} = e^{-m} = e^{-0.07} = 0.8187 \text{ (from the table)}$$

$$N\,(P_0) = (P_0) \times 10{,}000 = .8187 \times 10{,}000 = 8187$$

$$N(P_1) = N(P_0) \times m = 8187 \times .2 = 1637.4$$

$$N(P_2) = N(P_1) \times \frac{m}{2} = 92.14 = \frac{.44}{2} = 92.14 \times .22 = 20.27$$

$$N(P_2) = N(P_1) \times \frac{m}{2} = 1637.4 \times \frac{2}{2} = 163.74$$

The approximate number of packets containing not more than 2 defective blades in a consignment of 10,000 packets is:

10,000 – [8187 + 1637.4 + 163.4] = [10,000 – 9988.14] = 11.86 or 12.

Example 5:

Suppose that manufactured product has 2 defects per unit of product inspected. Using Poisson distribution, calculate the probabilities of finding a product without any defect, 3 defects and 4 defects. (Given $e^{-2} = 0.135$)

Solution:

Average number of defects of m = 2

$$P(r) = p(r) = \frac{e^{-m} m^r}{r!}, r = 0, 1,2,.....$$

$P(0) = e^{-2} = 0.135$ (given)^x

$P(1) = (P_0) \times m = .135 \times 2 = .27$

$P(2) = (P_1) \times m/2 = .27 \times 2/2 = .27$

$P(3) = (P_2) \times m/3 = .27 \times 2/3 = .18$

$P(4) = (P_3) \times m/4 = .18 \times .5 = .09$

Hence the probability that a product has no defect is 0.135, product has 3 defects is 0.18 and product has 4 defects is 0.09.

Poisson Distribution as an Approximation of the Binomial Distribution

The Poisson distribution can be a reasonable approximation of the binomial under certain conditions like:

(i) number of trials, *i.e.*, n is indefinitely large, *i.e.* $n \rightarrow \infty$.

(ii) p, *i.e.*, the probability of success for each trial is indenfinitely small,

(iii) np = = m (say) is finite.

The rule most often used by statisticians is that the Poisson is a good approximation of the binomial when n is equal to or greater than 20 and p is equal to or less than .05.

If above conditions hold good, we can substitute the mean of the binomial distribution (np) in place of the mean of the Poisson distribution (m), so that the formula becomes:

$$p(r) = \frac{e^{-np} (np)^r}{r!}$$

Proof:

In case of binomial distribution the probability of r successes is given by

$$p^{(r)} = {}^nC_r q^{n-r} p^r = \frac{n(n-1)......(n-r+1)}{r!} p^r q^{n-r}$$

Put $p = \frac{m}{n}$ $\quad\therefore q = 1 - p = 1 - \frac{m}{n}$

We now get $\quad p(r) = \frac{n(n-1\ldots\ldots n-r+1)}{r!}\left(\frac{m}{n}\right)^r \times \left(1-\frac{m}{n}\right)^{n-r}$

$$= \frac{1\left(1-\frac{1}{n}\right)\left(1-\frac{2}{n}\right)\ldots\ldots\left(1-\frac{r-1}{n}\right)m^r}{r!}\left\{\frac{\left(1-\frac{m}{n}\right)^n}{\left(1-\frac{m}{n}\right)^r}\right\}$$

For fixed r as $n \to \infty$

$\left(1-\frac{1}{n}\right)\ldots\left(1-\frac{r-1}{n}\right)\left(1-\frac{m}{n}\right)^r$ all tend to 1 and $\left(1-\frac{m}{n}\right)^n$ to $e^{-m}m^r$.

Hence in the limiting case $p^{(r)} = \frac{e^{-m}\, m^r}{r!}$

Needless to point out that use of Poisson as an approximation to binomial probability distribution if the conditions given above are satisfied can simplify calculation work and save time with more or less similar results as one would expect from binomial distribution.

HYPERGEOMETRIC DISTRIBUTION

The hypergeometric distribution occupies a place of great significance in statistical theory. It applies to sampling without replacement from a finite population whose elements can be classified into two categories—one which possesses a certain characteristic and another which does not possess that characteristic. The categories could be—male, female, employed, unemployed, etc. When n random selections are made without replacement from the population, each subsequent draw is dependent and the probability of success changes in each draw. The following conditions characterise the hypergeometric distribution:

(1) The result of each draw can be classified into one of two categories;

(2) The probability of a success changes on each draw;

(3) Successive draws are dependent; and

(4) The drawing is repeated a fixed number of times

The hypergeometric distribution which gives the probability of r successes in a random sample of n elements drawn without replacement is:

$$p(r)\ \frac{\binom{n-X}{b-r}(X_r)}{\binom{N}{n}} \quad \text{for } r = 0, 1, 2, \ldots, [N\ X].$$

Both the binomial distribution and the hypergeometric distribution are concerned with the same thing, viz., number of successes in a sample containing n observations. What differentiates these two discrete probability distributions is the manner in which data are obtained. For the binomial model, the sample data are drawn with replacement from a finite population or without replacement from an infinite population. On the other hand, for the hypergeometric model, the sample data are drawn without replacement from a finite population. Hence, while the probability of success p is constant over all observations of a binomial experiment and the outcome of any particular hypergeometric experiment; here the outcome of one observation is affected by the outcomes of the previous observations.

The symbol [n, X] means the smaller of n or X. The hypergeometric distribution bears a very interesting relationship to the binomial distribution. When N increases without limit, the hypergeometric distribution approaches the binomial distribution. Hence, the binomial probabilities may be used as approximation to hypergeometric probabilities where σ/N is small. A frequently used rule of thumb is that the population size should be at least ten times the sample size ($N > 10n$) for the approximation to be used.

Example:

A bag contains 20 balls of which 15 are of red colour and 5 of black colour. A random sample (without replacement) of 5 balls is taken. Find the probability that the sample contains 2 black balls.

Solution:

The problem can be solved with the help of hypergeometric probability model. Here N = 20, n = 5, r = 2. Hence the required probability is given by:

$$\frac{5_{C_2} \cdot 15_{C_3}}{39_C} = 0.29$$

NORMAL DISTRIBUTION

The binomial and the Poisson distributions described above are the most useful theoretical distributions for discrete variables, *i.e.*, they relate to the occurrence of distinct events. In order to have mathematical distribution suitable for dealing with quantities whose magnitude is continuously variable, a continuous distribution is needed. The *normal distribution,* also called the *normal probability distribution* happens to be most useful theoretical distribution for continuous variables. Many statistical data concerning business and economic problems are displayed in the form of normal distribution. In fact normal distribution is the *cornerstone* of modern statistics.

The normal distribution is an *approximation* to binomial distribution. Whether or not p is equal to q, the binomial distribution tends to the form of the continuous curve and when n becomes large at least for the material part of the range. As a matter of fact, the correspondence between the binomial and the curve is surprisingly close even for comparatively low values of n, provided that p and q are fairly near equality. The limiting frequency curve obtained as n becomes large is called the normal frequency curve or simply the normal curve.

The normal curve is represented in several forms. The following is the basic form relating to the curve with mean μ and standard deviation σ.

The Normal Distribution

$$P(X) = \frac{1}{\sigma\sqrt{2\pi}} e^{\frac{-(x-\mu)}{2\sigma^2}}$$

X = Values of the continuous random variable

μ = Mean of the normal random variable

e = Mathematical constant approximated by 2.7183

π = Mathematical constant approximated by 3.1416

$$\left(\sqrt{2}\pi = 2.5066\right)$$

When we say that the curve has unit area, we mean that the total frequency N is equated to i for convenience in representation and calculation. To obtain ordinates for a particular distribution the ordinates given by the above formula are multiplied by N. The equation to a normal curve corresponding to a particular distribution is thus given by

$$y = \frac{N}{\sigma\sqrt{2\pi}} e^{-x^2/2\sigma^2}$$

The quantity $\frac{N}{\sigma\sqrt{2\pi}}$ in the above formula is equal to the maximum ordinate (y_0) of the normal curve corresponding to distribution of stated total frequency N and stated standard deviation σ.

Graph of Normal Distribution

Remarks

(1) Normal distribution is a limiting case of binomial distribution when (i) $n \to \infty$ and (ii) neither p nor q is very small.

(2) The normal distribution can have different shapes depending on different values of μ and σ but there is one and only one normal distribution for any given pair of values of μ and σ.

(3) The two tails of the normal probability distribution extend indenfinitely and never touch the horizontal axis (which implies a positive probability for finding values of the random variable within any range from minus infinity to plus infinity).

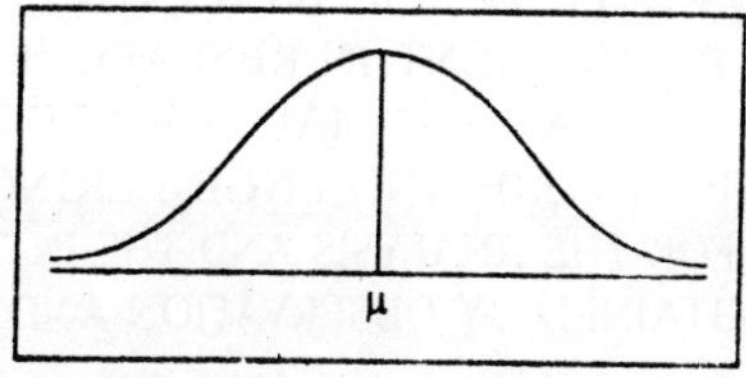

(4) Normal distribution is a limiting case of Poisson distribution when its mean m is large.

(5) The mean of a normally distributed population lies at the centre of its normal curve.

Relation between Binomial, Poisson and Normal Distributions

The three distributions, namely, Binomial, Poisson and Normal, are very closely related to each other. As explained earlier when N is large while the probability P of the occurrence of an event is close to zero so that q = (1 – p) the binomial distribution is very closely approximated by the Poisson distribution with m = np.

Since there is a relation between the binomial and normal distributions, it follows that there is also relation between the Poisson and normal distributions. In fact it can be proved that the Poisson distribution approaches a normal distribution with standardized variable $\frac{(x-m)}{\sqrt{m}}$ as m increases to infinity.

Importance of the Normal Distribution

The normal distribution has been long occupied a central place in the theory of Statistics. Its importance will be clear from the following points:

(1) The Normal distribution is used extensively in statistical quality control in industry in setting up of control limits.

A contemporary statistician W.J. Youden, whose hobby is typography, expresses his admiration of the normal distribution as follows:

THE
NORMAL
LAW OF ERROR
STANDS OUR IN THE
EXPERIENCE OF MANKIND
AS ONE OF THE BROADEST
GENERALIZATIONS OF NATURAL
PHILOSOPHY. IT SERVES AS THE
GUIDING INSTRUMENT IN RESEARCHES
IN THE PHYSICAL AND SOCIAL SCIENCE AND
IN MEDICINE, AGRICULTURE AND ENGINEERING. IT IS AN
INDISPENSABLE TOOL FOR THE ANALYSIS AND THE INTERPRETATION
OF THE BASIC DATA OBTAINED BY OBSERVATION AND EXPERIMENT.

Artistically enough it gives us the shape of the normal curve also.

(2) The normal distribution has numerous mathematical properties which make it popular and comparatively easy to manipulate. For example, the moments of the normal distribution are expressed in a simple form. The normal curve is reasonably close to many distributions of the humped type. If therefore we are ignorant of the exact nature of a humped distribution, or know the form but find it mathematically intractable, we may assume as a first approximation that the distribution is normal and see where this assumption leads us to.

(3) The normal distribution has the remarkable property stated in the so-called *Central Limit Theorem.* According to this theorem as the sample size n increases the distribution of mean, $\overline{X}$ of a random sample taken from practically any population approaches a normal distribution (with mean μ and standard deviation $\sigma/\sqrt{n}$) . Thus, if samples of large size, n, are drawn from a population that is not normally distributed, nevertheless the successive sample means will form themselves a distribution that is approximately normal. As the size of the sample is increased the sample means will tend to be normally distributed.

The central limit theorem applies to the distribution of most other statistics as well, such as the median and standard deviation (but not range). The central limit theorem gives the normal distribution its central place in the theory of sampling, since many important problems can be solved by this single pattern of sampling variability. As a result the work on statistical inferences in made simpler.

(4) In theoretical statistics many problems can be solved only under the assumption of a normal population. In applied work as well, we often find that methods developed under the normal probability law yield satisfactory results, even when the assumption of a normal population is not fully met, despite the fact that the problem can have a formal solution only if such a premise is hypothesized.

(5) As n becomes large the normal distribution serves as a good approximation of many discrete distributions (such as the Binomial or the Poisson model) whenever the exact discrete probability is laborious to obtain or impossible to calculate accurately.

Properties of the Normal Distribution

The following the important properties of the normal curve and the normal distribution.

1. There is one maximum point of the normal curve which occurs at the mean. The height of the curve declines as we go in either direction from the mean. The curve approaches nearer and nearer to the base but it never touches it, *i.e.*, the curve is asymptotic to the base on either side. Hence its range is unlimited or infinite in both directions.
2. The height of the normal curve is at its maximum at the mean. Hence the mean and mode of the normal distribution coincide. Thus for a normal distribution mean, median and mode are all equal.
3. The normal curve is "bell-shaped" and symmetrical in its appearance. If the curves were folded along its vertical axis, the two halves would coincide. The number of cases below the mean in a normal distribution is equal to the number of cases above the mean, which make the mean and median concide. The height of the curve for a positive deviation of 3 units is the same as the height of the curve for negative deviation of 3 units.
4. The points of inflexion, *i.e.*, the points where the change in curvature occurs are $\overline{X} \pm \sigma$.
5. Since there is only one maximum point, the normal curve is unimodal, *i.e.*, it has only one mode.

6. The area under the normal curve distributed as follows:
   (a) Mean ± 1 σ covers 68.27% area; 34.135% area will lie on either side of the mean.
   (b) Mean ± 2 σ covers 95.45% area.
   (c) Mean ± 3 σ covers 99.73% area.

The following table shows the area of the normal curve between mean ordinate and ordinates at various sigma distances from the mean as percentage of the total area.

Area Relationship

Distance from the mean Ordinate	**Percentage of Total Area**
0.5 σ	19.146
0.0 σ	34.134
0.5 σ	43.319
1.96 σ	47.500
2.0 σ	47.725
2.5 σ	49.379
2.5758 σ	49.500
3.0 σ	49.865

Thus the two ordinates at distance 1.96 σ from the mean on either side would enclose 47.5 + 47.5 = 95% of the total area, and two ordinates at 2.5758 σ distance from the mean on either side would enclose 49.5 + 49.5 = 99% of the total area. The area enclosed between ordinates at 3 σ distance from the mean on either side would be 49.865 + 49.865 = 99.73% of the total area. The various hypotheses are generally tested either at 5% level or at 1% level (*i.e.*, taking into account 95% and 99% of the total area of the normal curve).

Conditions for Normality

This following four conditions must prevail among the factors affecting the individual events that make up a given population, if the distribution of observations is to be normal:

(1) The causal forces must be numerous and of approximately equal weight.

(2) These forces must be the same over the universe from which the observations are drawn (although their incidence ill vary from event to event). This is the condition of homogeneity.

(3) The forces affecting events must be independent of one another.

(4) The operation of the causal forces must be such that deviation above the population mean are balanced as to magnitude and number by deviations below the mean. This is the condition of symmetry.

Constants of the Normal Distribution

The mean of the normal distribution is $\overline{X}$

The standard deviation of the normal distribution is σ

$$\mu_2 = \sigma^2,\ \mu^3 = 0 \text{ and } \mu_4 = 3\sigma^4$$

β_1 or moment coefficient of skewness

$$\beta_1 = \frac{\mu_3^2}{\mu_2^3} = 0$$

β_2 or moment coefficient of kurtosis

$$\beta_2 = \frac{\mu_4}{\mu_2^2} = \frac{3\sigma^4}{\sigma^4} = 3$$

For a normal distribution, the value of β_2 shall be 3. If the value of β_2 is more than 3, the curve is Leptokurtic and if the value of β_2 is less than three, the curve is platykurtic.

Area Under the Normal Curve

The equation of the normal curve gives the ordinate of the curve corresponding to any given value of x. However we are usually interested in areas under the normal curve instead of its ordinate. The area under the curve gives us the proportion of the cases falling between two numbers of the probability of getting a value between two numbers. The areas under the normal curve are tabulated and are shown in an appendix of tables given at the end of the book.

Before discussing the use of the table it is necessary to understand the meaning of the normal curve in its *standard form.* The equation of the normal curve depends on $\overline{X}$ and σ and for different values of $\overline{X}$ and σ we will obtain different curves. This would necessitate separate tables of normal curve areas for each pair if values of $\overline{X}$ and μ and an infinite number of tables would be example, $z = 1.8$ means that value of X of 1.8 σ to the right of μ. Similarly z = 2.2 means that the particular X value is 2.2. σ to the left of μ.

Since the various frequencies with which the values of a variable occur in the population can be expressed as populations by taking

$$P = \frac{F}{N}$$

and since $$\Sigma P = \frac{\Sigma P}{N} = \frac{N}{N} = 1$$

we can make the area under the curve corresponding to a normal distribution equal to unity regardless of the particular number of observations involved. We thus have a normal distribution that is independent of N, $\overline{X}$ and σ. The normal distribution in this form is called the *unit normal distribution,* or standard normal distribution and is applicable to any distribution that is normal in form regardless of the particular mean standard deviation and number of observations in the distribution. This theoretical distribution can be represented by a curve and the equation for the unit normal curve is

$$y = \frac{1}{\sigma\sqrt{2\pi}} e^{-z^2/2}$$

and $$z = \left(\frac{X - \overline{X}}{\sigma}\right)$$

The area under this curve is equal to 1. The curve is also called the *standard probability curve.* As X increases more and more, y becomes smaller approaches the horizontal axis but never touches it.

In any problem in which we are interested in determining area under the normal curve whose $\overline{X}$ and σ are 0 and 1, we only change the x's to z's and then use the table given at the end of the book. This table contains the normal curve area shaded in the following figure. In other words, the entries in the table equal the areas under the normal curve between the mean (z = 0) and the given value of z as shown below:

Entries corresponding to negative values of z are unnecessary because the normal curve is symmetrical. As a result the probability that z is between, say, –1.2 and zero, is equal to the probability that z lies between zero and 1.2. In other words, p $(0 \le z \le 1.20)$ = p $(-1.20 \le z \le 0)$.

The following are some of the examples to illustrate how tables are to be consulted in order to obtain area under the normal curve required. Fortunately the problem is solved by *standardizing* the data and only one table will be needed. We will be able to determine normal curve areas regardless of $\overline{X}$ and μ by tabulating only the area under the normal curve having $\overline{X} = 0$ and $\sigma = 1$.

Such *a normal curve with 0 mean and unit standard deviation is known as the standard normal curve.*

A normal curve with mean $\overline{X}$ and standard deviation σ can be converted into a standard normal distribution by performing the change of the scale and origin as indicated above. In the original scale (the x-scale) the mean and standard deviation are $\overline{X}$ and σ; in the new scale (the z-scale) they are 0 and 1. The formula that enables us to change form x-scale to z-scale and *vice versa is:*

$$z = \frac{X - \overline{X}}{\sigma} \text{ or } \frac{x}{\sigma}$$

where $x = (X - \overline{X})$.

In practice no matter what units of measurement the normal random variable x has (kgs., rupees, cms. hours, etc.) and no matter what combination of parameters μ_x and σ_x the data possess, we will always be able to convert the data onto a standardized scale by using the transformation formula and then determine the desired probabilities from the table of the standardized normal distribution.

This transformation from X to z is named as z *transformation* and has the effect of reducing X to units in terms of standard deviation.

Example 1:

Assume the mean height of soldiers to be 68.22 inches with a variance of 10.8 inches. How many soldiers in a regiment of 1,000, would you expect is be over six feet tall?

Solution:

Assuming that the distribution of height is normal.

Standard normal variate of z = $z = \dfrac{X - \overline{X}}{\sigma}$

Here $X = 72$ inches

$\overline{X} = 68.22$

$\sigma = \sqrt{10.8} = 3.286$

$$z = \frac{72 - 68.22}{3.286} = 1.15$$

Area to the right of the ordinate at 1.15 from the normal table is (0.5000 – 0.3749) = 0.1251. Hence the probability of getting soldiers above six feet is 0.1251 and out of 1,000 soldiers the expectation is 0.1251 × 1000 or 125.1 or 1.25. Thus the expected number of soldiers over six feet tall = 125.

Example 2:

A normal curve has $\overline{X} = 20$ *and* $\sigma = 10$. *Find the area between* $x_1 = 15$ *and* $x_2 = 40$.

Solution:

$$z_1 = \frac{X - \overline{X}}{\sigma} = \frac{15 - 20}{10} = -0.50$$

$$z_2 = \frac{40 - 20}{10} = +2.0.$$

Consulting the table we find the areas corresponding to the z's are 0.1915 and 0.4772 and thus the desire area between $x_1 = 15$ and $x_2 = 40$ is (0.1915 + 0.4772) = 0.6687 as shown below.

Significance of the Normal Distribution

The normal distribution is mostly used for the following purposes :

1. To approximate the distribution of means and certain other quantities calculated from samples, especially large samples.
2. To approximate of "fit" a distribution of measurement under certain conditions.
3. To approximate the binomial distribution and other discrete of continuous probability distributions under suitable conditions.

The following examples shall illustrate the applications of normal distribution:

Example 3:

1,000 light bulbs with a mean life of 120 days are installed in new factory, their length of life is normally distributed with standard deviation 20 days (i) How many bulbs will expire in less than 90 days? (ii) If it is decided to replace all the bulbs together, what intervals should be allowed between replacements if not more than 10 percent should expire before replacement?

Solution:

(i) $\overline{X} = 120, \sigma = 20, X = 90$

Standard normal variate or

$$z = \frac{90 - 120}{20} = -1.5$$

Area of the curve (z = – 1.5) up to the mean ordinate – 0.4332

Area of the left of – 1.5 = 0.5 – 0.4332 = 0.0668.

Example 4:

In a distribution exactly normal, 7% of the items are under 35 and 79% are under 63. What is the mean and standard deviation of the distribution?

Solution:

Since 7% of items are under 35, 43% are between $\overline{X}$ and 35. Similarly the percentage of items between $\overline{X}$ and 63 is 39%.

The standard normal variate corresponding to 0.43 (43%) is 1.48. Thus

$$\frac{35 - \overline{X}}{\sigma} = -1.48$$

The standard normal variate corresponding to 0.39 (39%) is 1.23

$$\frac{63 - \overline{X}}{\sigma} = +1.23$$

From (i) and (ii)

$$1.48\ \sigma + \overline{X} = -35$$

$$1.23\ \sigma + \overline{X} = 63$$

On adding these equations, we get

$$271\ \sigma = 28$$

$$\sigma = \frac{28}{2.71} = 10.33$$

$$1.46 \times 10.33 - \overline{X} = -35$$

$$-\overline{X} = -35 - 15.3$$

$$\therefore \quad -\overline{X} = 50.3.$$

Hence the mean of the distribution is 50.3 and σ = 10.33.

MISCELLENEOUS EXAMPLES

Example 1:

The Bombay Municipal Corporation installed 2,000 bulbs in the streets of Bombay. If these bulbs have an average life of 1,000 burning hours, with a standard deviation of 200 hours, what number of bulbs might be expected to fail in the first 700 burning hours? The table of area of the normal curve at selected values is as follows :

$\frac{X - \overline{X}}{\sigma}$	*Probability*
1.00	*0.159*
12.5	*0.106*
1.50	*0.067*

Solution:

Average lie of bulbs $\left(\overline{X}\right)$ = 1,000 hours

$$\sigma = 200 \text{ hours}$$

$$X = \text{buming hours} = 700$$

$$z = \frac{X - \overline{X}}{\sigma} = \frac{700 - 1{,}000}{200} = -1.5$$

Area to the left of (– 1.5) = 0.067.

∴ Number of bulbs expected to fail in the first 700 hours

$$= 0.067 \times 2{,}000 = 134.$$

Number of bulbs expected to expire in less than 90 days

$$= 0.0668 \times 1{,}000 = 66.8 \text{ or } 67.$$

(ii) The value of standard normal variate corresponding to an area 0.4 (0.5 – 0.1) is 1.28.

$$\frac{X - 120}{20} = -128$$

$$X = 120 - (1.28 \times 20)$$

$$= 120 - 25.6 = 94.4 \text{ or } 94.$$

Hence the bulbs will have to be replaced after 94 days.

Example 2:

In an intelligence test administered to 1,000 students the average score was 42 and standard deviation 24. Find (a) the number of students exceeding a score of 50, (b) the number of students lying between 30 and 54, (c) the value of score exceeded by the top 100 students.

Solution:

(a) Given $\overline{X} = 42$, $X = 50$, $s = 24$

$$z = \frac{X - \overline{X}}{\sigma} = \frac{50 - 42}{24} = 0.333$$

Area to the right of ordinate at 0.333 is 0.5 – 0.1304 = 0.3696

∴ The expected number of children exceeding a score of 50

= 0.3696 × 1,000 = 369.6 or 370.

(b) Standard normal variate for score 30

$$z = \frac{X - \overline{X}}{\sigma} = \frac{30 - 42}{24} = -0.5$$

Standard normal variate for score 54

$$z = \frac{X - \overline{X}}{\sigma} = \frac{54 - 42}{24} = 0.5$$

Area from z = 0 to z = 0.5 = 0.1915

Area from z = – 0.5 to z = 0= 0.1915

∴ Area from z = – 0.5 to z = 0.5 = 0.1915 + 0.1915

= 0.3830

Thus the number of children having score between 30 and 54

= 0.383 × 1,000 = 383.

(c) Probability of getting top 100 students

$$= \frac{100}{1000} = 0.1$$

Standard normal variate having 0.1 are to the right = 2.81

Standard normal variate for score X

$$z = \frac{X - \overline{X}}{\sigma}$$

$$1.281 = \frac{X - 42}{24}$$

$$1.28 \times 24 = X - 42$$

$$X = (1.28 \times 24) + 42 = 72.72 \text{ or } 73.$$

Example 3:

A sample of 100 dry battery cells tested to find the length of life produced the following results :

$\overline{X}$ = *12 hours,* σ = *3 hours*

Assuming the data to be normally distributed, what percentage of battery cells are expected to have life :

(i) More than 15 hours,

(ii) Less than 6 hours, and

(iii) Between 10 and 14 hours?

[Given z :	*2.5*	*2*	*1*	*0.67*
Area	*0.4938*	*0.4772*	*0.3413*	*0.2487*

Solution:

Let the random variable X denote the length of life of dry battery cells.

(i) $P(X = 15) = P(Z = 1)$

where $Z = \dfrac{15-12}{3} = 1$

Area under the standard normal curve to the right of Z = + 1

= 0.05 – 0.3413 = 0.1587 or 15.87%

(ii) $P(X < 6) = P(Z < -2)$

where $Z = \dfrac{6-12}{3} = -2$.

Area under the standard normal curve to the left to Z = – 2

= Area to the right of Z = + 2

= 0.5 – 0.4772 = 0.228 = 2.28%

(iii) $P(10 < X < 14) = P(-0.067 < Z < 0.67)$

Area under the standard normal curve between Z = – 0.67 to Z = 0.67

= 2 (Area between Z = 0 and Z = 0.67)

= 2 × 0.2487 = 0.4974 = 49.74%.

Example 4:

A project yields an average cash-flow of Rs. 500 lakhs with a standard deviation of Rs. 60 lakhs Calculate the following probabilities :

(i) Cash-flow will be more than Rs. 560 lakhs

(ii) Cash-flow will be less than Rs. 420 lakhs

(iii) Cash-flow will be between Rs. 460 lakhs and Rs. 540 lakhs.

(iv) Cash-flow will be more than Rs. 680 lakhs.

(i) Standard normal variate corresponding to 420, *i.e.*,

$$Z = \frac{460 - 500}{60} = -1.33$$

Area to the left of the ordinate at – 1.33 = 0.5 – 0.4082 = 0.918

(iii) Standard normal variate corresponding to 460, *i.e.*,

$$Z = \frac{460 - 500}{60} = -0.67$$

Standard normal variate corresponding to 540 = 540 – 500

Area between Z = – 0.67 and Z = + 0.67 = 0.2486 + 0.2486 = .4972

The probability that cash flow will be between 460 lakhs and Rs. 540 lakhs = 0.4572

(iv) Standard normal variate corresponding to 680, *i.e.*,

$$Z = \frac{680 - 500}{60} = 3$$

Area to the right of the ordinate at

3 = 0.5 – 0.4987 = 0.0013.

Example 5:

If 20% bolts produced by a machine are defective, determine the probability that out of 4 bolts chosen at random (a) 1 (b) 0, (c) at most 2 bolts will be defective.

Solution:

Probability of defective bolt = p = 0.2

Probability of a non-defective bolt = 1 – 0.2 = 0.8

(a) Probability (1 defective bolt) = $^4C_1\ (.2)^2\ (.8)^3$ = 0.4096

(b) Probability (0 defective bolts) = $^4C_2(0.2)^0\ (0.8)^4$ = 0.4096

(c) Probability (2 defective bolts) = $^4C_2(0.2)^2\ (0.8)^2$ = 0.1536

pr. (at most 2 defective bolts) = pr. (0 def. bolts) + pr. (1 def. bolt) + pr. (2 def. bolts)

= 0.4096 + 0.4096 + 0.1536 = 0.9728.

Example 6:

Out of 320 families with 5 children each, what percentage would be expected to have (i) 2 boys and 3 girls, (ii) at least one boy? Assume equal probability for boys and girls

Solution:

(i) $p = q = \frac{1}{2}$

The percentage of families expected to have 2 boys and 3 girls shall be given by the binomial distribution

$$^5C_2\, p^2 q^3 = {^5C_2}\left(\frac{1}{2}\right)^2\left(\frac{1}{2}\right)^3 = \frac{5 \times 4}{2} \times \frac{1}{32} = \frac{5}{16} \text{ or } 31.25\%$$

(ii) Probability of getting at least one boy shall be given by:

$$1 - {^5C_5}\, q^5 = 1 - \frac{1}{32} = \frac{31}{32} \text{ or 97 per cent.}$$

Example 7:

A binomial distribution has n = 20 and p = 3. Find the mean and the variance of this distribution.

Solution:

The mean of binomial distributions np and variance npq.

$n = 20 \; p = 3$

Hence mean, *i.e.*, $np = 20 \times .3 = 6$

Varlance = $npq - 20 \times .3 \times .7 = 4.2$

Example 8:

A manufacturer who produces medicine bottles, finds that 0.1% of the bottles are defective. The bottles are packed in boxes containing 500 bottles. A drug manufacturer buys 100 boxes from the producer of bottles. Using Poisson distribution, find how many boxes will contain :

(i) no defectives.

(ii) at least two defectives.

(Given $e^{0.5} = 0.6065$)

Solution:

We are given $p = 0.001$ and $n = .500$; $m = np = 0.5$, $N = 100$

(i) No. of boxes that will contain no defectives :

$$= NP_{(0)} = N\frac{e^{-m}\ m^r}{r!}$$

$$= 100 \ ' \ e^{-0.5} = 100 \times .6065 = 60.65 \text{ or } 61.$$

(ii) No. of boxes that will contain at least two defectives

$$= N[1 - P(0) - P(1)] \qquad = 100[1 - e^{-0.05} - e^{-0.05}]$$

$$= 100[1 - .6065 - .30325] = 100\ (.09025) = 9.025 \text{ or } 9$$

Example 9:

In Deli with 100 municipal wards, each having approximately the same population, the distribution of typhoid cases in 1987 was as follows:

No. of Cases :	*0*	*1*	*2*	*3*	*4*
No. of Wards :	*63*	*28*	*6*	*2*	*1*

Fit a Poisson distribution for the above.

Solution:

Fitting of Poisson Distribution

X	f	fX
0	63	0
1	28	28
2	6	12
3	2	6
4	1	4
	N = 100	*S fX = 50*

$$\overline{X} = \frac{\Sigma fX}{N} = \frac{50}{100} = \frac{1}{2}$$

$$P_{(0)} = e^{-m} = e^{-0.5} = 0.6065 \qquad \text{(from the table)}$$

$$N(P_0) = P_{(0)} \times N = .6065 \times 100 = 60.65 \approx 30$$

$$N(P_1) = N(P_0) \times m = 60.65 \times \frac{1}{2} \ 30.33 \approx 30$$

$$N(P_2) = N(P_1) \times \frac{m}{2} = 30.33 \times \frac{1}{4} = 7.58 \approx 8$$

$$N(P_3) = N(P_2) \times \frac{m}{3} = 7.58 \times \frac{1}{6} \ 1.26 \approx 1$$

$$N(P_4) = N(P_3) \times \frac{m}{4} = 1.26 \times \frac{1}{4} = 0.16 \approx 0.$$

Example 10:

Of a large group of men, 4 per cent are under 60 inches in height and 40 per cent are between 60 and 65 inches. Assuming a normal distribution, find the mean height and standard deviation.

Solution:

Let the random variable X represent the height of men and $\overline{X}$ and s the mean and standard deviation of the normal distribution.

The area of the Left of X = 60 – 0.05

The area between Z = 0 and $Z_1 = \frac{X-60}{\sigma}$ = 0.5 – 0.05 = 0.45

From the table the value of Z_1 corresponding to this area is 1.645. Hence

$$\frac{60-\overline{X}}{\sigma} = -1.645 \qquad \text{...(i)}$$

The height of 40% of the mean between 60 and 65 inches is the area between the ordinate of X = 60 and X = 65 is 0.4 or the left of the ordinate at X=65=0.45. The area between Z = 0 and $Z_2 = \frac{65-\overline{X}}{\sigma}$ is 0.50 – 0.45=0.05.

The value of Z_2 corresponding to this area is 0.13.

$$\therefore \qquad \frac{65-\overline{X}}{\sigma} = -0.13$$

Dividing Eqn. (i) by (ii), we get $\frac{60-\overline{X}}{65-X} = \frac{1.645}{0.13}$ or $\overline{X}$ = 65.42

Putting the value of $\overline{X}$ in Eqn. (i) $\sigma = \frac{60-65.42}{1.645} = 2.29$

Hence $\overline{X}$ = 65.42 and σ = 3.29.

Example 11:

Five air coins were tossed 100 times. From the following outcomes, calculate expected frequencies :

No. of heads up :	*0*	*1*	*2*	*3*	*4*	*5*
Observed frequency :	*2*	*10*	*24*	*35*	*18*	*8*

Solution:

Probability of head = $\frac{1}{2}$, Probability of tail = $\frac{1}{2}$. The expected Binomial frequencies can be obtained by expanding :

$$100\left(\frac{1}{2}+\frac{1}{2}\right)^5$$

$$(q + p)^n + q^n\ {}^nC_1\ q^{n-1}\ p + {}^nC_2\ q^{n-2}\ p^2 + \ldots$$

$$=100\left[\left(\frac{1}{2}\right)^5+\left(\frac{1}{2}\right)^4\left(\frac{1}{2}\right)+10\left(\frac{1}{2}\right)^2\left(\frac{1}{2}\right)^2+10\left(\frac{1}{2}\right)^2\left(\frac{1}{2}\right)^3+5\left(\frac{1}{2}\right)\left(\frac{1}{2}\right)^4+\left(\frac{1}{2}\right)^5\right]$$

$$= 100\left(\frac{1}{32}+\frac{5}{32}+\frac{10}{32}+\frac{5}{32}+\frac{1}{32}\right) = 3.125,\ 15.625,\ 31.25,\ 15.625,\ 3.125.$$

Example 12:

There are 600 business students in the post-graduate department of a university, and the probability for any student to need a copy of a particular textbook from the university library on any day is 0.05. How many copies of the book should be kept in the university library so that the probability may be greater than 0.90 that none of the students needing a copy from the library has to come back disappointed. (Use normal approximation to the binomial probability law.)

Solution:

Let n be the number of students and p the probability for an student to need a copy of a particular textbook from the university library.

Mean : $\overline{X} = np = 600 \times .05 = 30$

$$\sigma = \sqrt{npq} = \sqrt{600 \times .05 \times .95} = 5.34$$

Let x_1 represent the number of copies of a textbook required on any day. We want x_1 such that

$P(X < z_1) > 0.9$ or $P(Z\ (z_1) > 0.90)$

$\left(z_1 = \dfrac{x_1 - 30}{5.34}\right)$ or $P(0 < Z(z_1) > 0.4$ or $z_1 > 1.28$ [From normal tables]

$$\therefore \frac{x_1 - \mu_1}{\sigma} > 1.28 \text{ or } \frac{x_1 - 30}{5.3} > 1.28$$

$x_1 - 30 > 7.784;$ $\quad x_1 > 36.784 = 37$

Hence the library should keep at least 37 copies of the book to ensure that the probability is more than 90% that none of the students reading a copy from the library has to come back disappointed.

Example 13:

The mean weight of 500 male students in a certain college is 151 lb. and the standard deviation is 15 lb. Assuming the weights are normally distributed find how many students weigh: (a) between 120 and 155 lb. and (b) more than 185 lb.

Solution:

(a) Weights recorded as being between 120 and 155 lb can actually have any value from 119.5 to 155.5 lb. assuming they are recorded to the nearest pound.

Standard normal variate corresponding to 119.5 lb.

$$z = \frac{119.5 - 151}{15} = -2.1$$

Standard normal variate corresponding to 155.5 lb.

$$= z = \frac{155.5 - 151}{15} = 0.3$$

Area between z = – 2.10 and z = 0.30

= 0.4821 + 0.1179 = 0.6,

∴ The number of students weighing between 120 and 155 lb.

= 500 × 0.6 = 300.

(b) Students weighing more than 185 lb. must weigh at least 185 lb.

Standard normal variate corresponding to 185.5

$$= z = \frac{185.5 - 151}{15} - 2.3$$

Area to the right z = 2.3 is (0.5000 – 0.4893) = 0.0107.

∴ Number of students weighing more than 185 lb.

= 500 × 0.01 × 07 = 5.35 or 5 app.

Example 14:

The income of a group of 10,000 persons was found to be normally distributed with mean = Rs. 750 p.m. and standard deviation = Rs. 50. Show that of this group about 95% had income exceeding Rs. 668 and only 5% had come exceeding Rs. 832. What was the lowest income among the richest 100?

Solution:

Standard normal variate or

$$z = \frac{\left(X - \overline{X}\right)}{\sigma}$$

Hence $X = 668, \overline{X} = 750, s = 50$

$$z = \frac{668 - 750}{50} = \frac{-82}{50} = -1.64$$

Area of the right of the ordinate at – 1.64 is (0.4495 + 0.5000) = 0.9495.

∴ The expected number of persons getting above Rs. 668
= 10,000 × 0.9495 = 9495

This is about 95% of the total *i.e.*, 10,000

The standard normal variate corresponding to 832 is

$$z = \frac{832 - 750}{50} = \frac{82}{50} = 1.64.$$

Area to the right of ordinate at 1.64 is

$$0.5000 - 0.4495 = 0.0505$$

The number of persons getting above Rs. 832 is

$$10{,}000 \times 0.0505 = 505$$

This is approximately 5%.

Probability of getting richest 100

$$= \frac{100}{10000} = 0.01.$$

Standard normal variate having 0.01 area to its right = 2.33

$$2.33 = \frac{X - 750}{50}$$

$$X = (2.33 \times 50) + 750 = \text{Rs. } 866.5.$$

Hence the lowest income of the richest 100 persons is Rs. 866.50.

Example 15:

A grinding machine is so set that its production of shafts has an average diameter of 10.10 cms and a standard deviation of 0.20 cms. The product specifications call for shaft diameters between 10.05 cms and 10.20 cms. What proportion of output meets the specifications presuming normal distribution?

Solution:

Assuming that the distribution of diameter of shafts is normal.

We have μ = 10.10 cm, σ = 0.20 cm.

For x = 10.05, $z = \frac{10.05 - 10.10}{0.20} = -0.25$

For x = 10.20, $z = \frac{10.20 - 10.10}{0.20} = 0.5$

$\therefore$ Area of the normal curve between the ordinate – 0.25 to 0.5 = Area between the ordinates – 0.25 to + Area between the ordinates 0 to 0.5 = 0.0987 + 0.1915 = 0.2902. Thus 29.02% of the output meets the specification.

Example 16:

As a result of tests on 20,000 electric bulbs manufactured by a company it was found that the lifetime of the bulb was normally distributed with an average life of 2,040 hours and standard deviation of 60 hours. On the basis of the information estimate the number of the bulbs that are expected to bum for (a) more than 2,150 hours, and (b) less than 1,960 hours.

Proportion of Area under the Normal Curve

z	*Area*	*z*	*Area*	*z*	*Area*
1.23	*0.3907*	*1.33*	*0.4082*	*1.43*	*0.4236*
1.63	*0.4484*	*1.73*	*0.4582*	*1.83*	*0.4664*

Solution:

(a) $\overline{X}$ = 2040, X = 2150, σ = 60

$$z = \frac{X - \overline{X}}{\sigma} = \frac{2150 - 2040}{60} = \frac{110}{60} = 1.833$$

Area of the right of ordinate at 1.833

0.5 – 0.4667 = 0.0333

The number of bulbs that are expected to burn more than 2,150 hours

0.0333 × 20,000 = 666

(b) X = 1960, $\overline{X}$ = 2040, 60

$$z = \frac{1960 - 2040}{60} = 1.333$$

Area to the left of ordinate at 1.333

= 0.5 – 0.4082 = 0.0918

$\therefore$ The number of bulbs that are expected to bum for less than 1,960 hours

$= 0.0198 \times 20{,}000 = 1{,}836.$

Example 17:

The results of a particular examination are given below in summary form:

Results	*% of candidates*
(i) Passed with distinction	*10*
(ii) Passed	*60*
(iii) Failed	*30*

It is known that a candidate gets plucked if he obtains less than 40 marks (out of 100) while he must obtain at least 75 marks in order to pass with distinction. Determine the mean and standard deviation of the distribution of marks assuming this to be normal.

Solution:

We have to calculate the mean and standard deviation from the given information.

The following diagram will help in understanding the question and finding its solution:

We know that 30% students get less than 40 marks.

∴ From the table the z value corresponding to

3.2 (20% area) = – 0.524 z = 0.52 for 0.1985

z = 0.53 for 0.2019

∴ z = 0.524 for 0.200

$$\text{Hence } \frac{40 - \overline{X}}{\sigma} = -0.524$$

10% students get distinction marks, *i.e.*, 75 or more.

From the table the z value corresponding to

0.4 (40% area) = 1.28.

$$\text{Hence } \frac{75 - \overline{X}}{\sigma} = 1.28.$$

From equations (i) and (ii)

$$\begin{aligned} \overline{X} - 40 &= 0.524\,\sigma \\ -\overline{X} + 75 &= 1.280\,\sigma \\ \hline 35 &= 1.804\,\sigma \end{aligned}$$

$$\sigma = \frac{35}{1.804} = 19.4$$

$$40 - \overline{X} = -0.524 \times 19.4$$

$$-\overline{X} = -10.7 - 40 \text{ or } \overline{X} = 50.17.$$

Hence the mean of the distribution is 50.17 and standard deviation 19.4.

Example 18:

(a) 15,000 students appeared for an examination. The mean marks were 49 and the standard deviation of marks was 6. Assuming the marks to be normally distributed, what proportion of students scored more than 55 marks?

(b) If in the same examination, Grade 'A' is to be given students scoring more than 70 marks what proportion of the students will receive Grade 'A'?

Solution:

(a) $z = \dfrac{X - \overline{X}}{\sigma}$

$X = 55, \overline{X} = 49, \sigma = 6$

$\therefore\ z = \dfrac{55 - 49}{\sigma} = 1$

The proportion of students scoring between 49 and 55 (or between zero and one on the standard scale) is 0.3413.

$\therefore$ The proportion of students scoring more than 55 marks is

$0.5 - 0.3413 = 0.1587$ or 15.87%.

(b) $z = \dfrac{X - \overline{X}}{\sigma} = \dfrac{70 - 49}{6} = \dfrac{21}{6} = 3.5$

The area under the standard normal curve corresponding to 3.5

= 0.4998.

Therefore, 0.02 per cent (0.5 – 0.4998) = 0.0002 would score more than 70 marks. Since there are 15,000 candidates, 3 candidates will receive Grade 'A'.

Fitting a Normal Curve

There are two main objects of fitting a normal curve to sample data:

1. To prove a visual device for judging whether or not the normal curve is a good fit to the sample data, and

2. To use the smoothed normal curve, instead of the irregular curve representing the sample data, to estimate the characteristics of the population.

Methods of Fitting

The following two methods are used to fit a normal curve to an observed set of data:

1. Method of ordinates, and
2. Method of areas.

Methods of Ordinates

In or to draw a curve on a graph, we need the frequencies which are represented on the ordinate and the values of the variable which are shown on the abscissa. Hence in order to fit a curve we must know the frequencies (ordinates) at the various of the abscissa scale. While fitting a normal curve the ordinates are obtained at various sigma distances from the mean. The procedure of obtaining ordinates is as follows:

The height of the mean ordinate is given by

$$-\frac{1}{2}\left(\frac{X-\overline{X}}{\sigma}\right)^{\frac{1}{2}}$$

$$y = \frac{Ni}{2.5066\ \sigma}\ 2.71828$$

We need the value of N, $\overline{X}$ and s in order to fit a normal curve in a distribution. In the expression above when $(X - \overline{X}) = 0$, the exponent of 2.71828 raised to the zero power is one. *This the expression $e^{-x^2\ 2\sigma^2}$ e always equal to 1 for the ordinate erected at the mean. The mean ordinates can easily be obtained by*

$$y_2 = \frac{Ni}{2.5066\ \sigma}.\ \text{or}\ y_3 = 0.399\left(\frac{Ni}{\sigma}\right)$$

y_2 denotes the ordinate to be erected at the mean and t denotes the width of the class interval.

This is the maximum ordinate of the fitted curve. The height of the ordinate at a distance 1 σ from the mean would be calculated as follows:

$$y_1 = 0.399\left(\frac{Ni}{\sigma}\right) e^{-\frac{1}{2}(1)^2}$$

In a similar manner the height of the ordinate at a distance of 2s from the mean would be calculated.

In practice it is not necessary to compute the ordinates at various sigma distances from the mean in the above manner. What is done is that only the height of the mean ordinate is calculated in the manner given above and the heights of other ordinates are found out from a specially prepared mathematical table which gives the ordinates of the Normal Probability Curve. (The table is given at the end of the book). In *order to consult the table the distance from standard normal variate, i.e. the point* $\left[\frac{X - \overline{X}}{\sigma}\right]$ *denoted as* $\frac{x}{\sigma}$ the corresponding value is read. This value gives the height of the ordinate as a proportion of the mean ordinate. *Therefore, to get a value of the ordinate at that point this has to be multiplied by the height of the mean ordinate.*

While fitting the normal curve the question is at what points on the horizontal scale should the ordinates be erected. Generally the ordinates are erected at class mid-points.

Example 19:

How would you use the normal distribution to find approximately the frequency of exactly 5 successes in 100 trials, the probability of success in each trial being p = 0.1.

Solution:

Let n = number of trials, p = probability of success and q = the probability of failure. Hence n = 100, p = 0.1 and q = 0.9. It is a case of binomial distribution. For binomial distribution:

$$\overline{X} = np \text{ and } q = \sqrt{npq}$$

Hence $\overline{X} = np = 100 \times .01 = 10$

$$\sigma = \sqrt{100 \times 0.1 \times 0.9} = 9$$

We know that when number of trials is large, binomial distribution tends to normal distribution. Normal distribution is a continuous distribution. Frequency of exactly 5 trials of binomial will correspond to the frequency of class interval 4.5 to 5.5 and standard deviation of binomial distribution will corresponded to mean and standard deviation of normal distribution.

$$= \frac{4.5 - np}{\sqrt{npq}} = \frac{4.5 - 10}{3\sigma} = -1.83$$

Standard normal variate correspond to 5.5

$$= \frac{5.5 - np}{\sqrt{npq}} = \frac{5.5 - 10}{3} = -1.50$$

Area below the value – 1.83 from the table = 0.0336

Area below the value – 1.50 from the table = 0.0668

Area between the two = 0.0668 – 0.0336 = 0.00332

The frequency of class interval 4.5 to 5.5

∴ Approximate frequency of exactly 5 successes in 100 trials with p = 0.1

= 0.0332 × 100 = 3.32.

Example 20:

Suppose that in key punching of 80 column IBM cards, the arithmetic mean number of mistakes per card is 0.3. What per cent of cards will have (i) no mistake, (ii) one mistake, and (iii) two mistakes.

Solution:

We are given m = 0.3

Probability of mistake, *i.e.*, no mistake = $e^{-m} = 2.7183^{-0.3}$

= Rec. [AL (Log 2.7183 × 0.3)]

= Rec. [Al (0.4343 × 0.3)]

= Rec. [AL 0.13029]

= Rec. 1.35 = 0.7408.

This means that 74 per cent of the cards punched will have no mistakes.

$P(1) = P_0 \times m = 0.7408 \times 0.3 = 0.22224.$

This means that about 22 per cent of the cards punched will have one mistake.

$$P_2 = P_1 \times \frac{m}{2} = 0.22224 \times \frac{0.3}{2} = 0.0333.$$

This means that about 3 per cent of the cards punched will have two mistakes.

Example 21:

Fit a normal curve to the following data by (a) the method of ordinates, and (b) the method of areas :

Variable	*Frequency*
60-62	*5*
63-65	*18*
66-68	*42*
69-71	*27*
72-74	*8.*

Solution:

(a) Fitting normal curve by the method of ordinates.

For fitting the normal curve we need the values of $\overline{X}$ and σ.

Calculation of $\overline{X}$ and s

Variable	m	f	d	fd	fd²
60-62	61	5	–2	–10	20
63-65	64	18	–1	–18	18
66-68	67	42	0	0	0
69-71	70	27	+1	+27	27
72-74	73	8	+2	+16	32
		N = 10		Σfd = 15	Σ fd² = 97

$$\overline{X} = A + \frac{\Sigma\, fd}{N} \times i = 67 + \frac{15}{100} \times 3 = 67.45$$

$$\sigma = \sqrt{\frac{\Sigma\, fd^2}{N} - \left(\frac{\Sigma\, fd}{N}\right)^2} \times i = \sqrt{\frac{97}{100} - \left(\frac{15}{100}\right)^2} \times 3$$

$$= \sqrt{0.97 - 0.0225} \times 3 = 0.973 \times 3 = 2.92.$$

$$\text{The mean ordinate} = \frac{0.39 \times 100 \times 3}{2.92} = 41.$$

Therefore, the height of the maximum ordinate at the point 67.45 on the abscissa will be 41. *Weight of the ordinate of 1 σ from the mean.*

This can be obtained either from the equation or from the specially prepared mathematical table which gives the ordinates of the normal probability curve. Use of the table is very convenient because it saves a lot of labour. From the table the proportion of ordinate at 1s from the mean = 0.60653.

∴ Height of the ordinate at 1s from the mean

$= 0.60563 \times 41 = 24.87$ or 25.

Similarly the height of the ordinate at 2σ from the mean

$= 0.13534 \times 41 = 5.55$

and the height of the ordinate at 3σ from the mean

$= 0.0111 \times 41 = 0.455$

With the help of the mean ordinate 41 and $\overline{X} = 1\sigma$, $\overline{X} = 2\sigma$, etc. the normal curve can easily be plotted on the graph paper. The points to be plotted on the left-hand side of mean ordinate are 0.455, 5.55 and 25 and the points on the right-hand side or mean ordinate are 25, 5.55 and 0.455.

The following table shows how to obtain the height of ordinates at mid-points of various class intervals :

The following table shows how to obtain the height of ordinates at mid-points of various class intervals :

Variable	Mid-points	$(X-\overline{X})$ x	$\frac{x}{\sigma}$	Proportionate ordinate, *i.e.*, $-\frac{1}{2}\left(\frac{x}{\sigma}\right)^2$	Height of ordinates or Expected frequency
(1)	(2)	(3)	(4)	(5)	(6)
60-62	61	–6.45	2.209	0.08679	3.56 or 4
63-65	64	–3.45	1.182	0.49750	20.40 or 20
66-68	67	–0.45	0.154	0.98820	40.52 or 41
69-71	70	+2.55	0.873	0.68313	28.01 or 28
72–74	73	+ 5.55	1.901	0.16448	6.74 or 7
					N = 100

Steps in Computing Height of Ordinates. The steps in the computation of heights of ordinates are :

1. Find the mid-points of the various classes. (Col. 2)
2. Take the deviation of each of the mid-points from the mean, *i.e.*, obtain $(X-\overline{X})$ and denote this column by x. (Col. 3.)
3. Divide each value of x by σ, *i.e.*, obtain $\frac{x}{\sigma}$. (Col. 4)

4. Calculate the height at distances corresponding to $\frac{x}{\sigma}$ from the table giving ordinates of the normal curve. (Col. 5).
5. Multiply each figure of Col. 5 by the value of mean ordinate height (41). The resultant figures are the values of the heights of ordinates at various distances from the mean, or the expected frequencies of the normal distribution at various class mid-points. (Col. 6)

Method of Areas : The area under the normal curve represents the total number of frequencies. Tables have been prepared which give areas under the normal curve. The tables give the proportion or percentage of area between an ordinate erected at the mean and an ordinate at a given distance from the mean in standard deviation units.

The mean of the distribution is located at the centre of the distribution and it cuts the curve into two equal parts. Mean ± 1σ on either side of the mean covers 68.27 per cent items, mean ± 2σ covers 95/45 per cent items and mean ± 3σ covers 99.73 per cent items. When the method of areas is used to fit the normal curve we obtain area (frequencies) *within* the various *class intervals.*

The method of areas has been applied to the problem of illustration 32 to obtain the expected frequencies of the normal distributions.

Fitting The Normal Curve by the Method of Areas

Class limits	Lower class limit	Observed frequency	(X–67.45) x	$\frac{x}{\sigma}$	Area under	Area for each class normal curve from 0 to z	Expected frequency
(1)	(2)	(3)	(4)	(5)	(6)	(7)	(8)
59.5–62.5	59.5	5	–7.95	–2.72	0.4967	0.0413	4.12 or 4
62.5–65.5	62.2	42	–1.95	–0.67	0.2486	0.3892	38.92 or 39
68.5–71.5	68.5	27	+1.05	+0.36	0.1406	0.2771	27.71 or 28
71.5–74.5	71.5	8	+4.05	+1.39	0.4177	0.0743	7.43 or 7
	74.5		+7.05	+2.41	0.4920		

Steps: The method of computing the expected frequencies is given below.

- Column (1) gives the class intervals of the series to which the normal curve is to be fitted.
- Column (2) gives the lower class limit of the various class intervals.
- Column (3) gives the actual or observed frequencies.
- Column (4 gives the deviations of the lower class limits from the mean, *i.e.*, $(X - \overline{X})$. Thus for class interval 59.5–62.5 we have the value (59.5–67.45) or –7.95, and so on.
- In Column (5) the deviations are expressed in units of standard deviation of the series.
- Column (6) has been derived from the table of areas under the normal curve. For example, corresponding to $\frac{x}{\sigma} = 2.72$ the area under the normal curve is 0.4967.
- Column (7) is the column of differences. It gives areas under the normal curve between successive values of z. These are obtained by subtracting the successive areas in the sixth column when the corresponding z's have the same sign, and adding them when the z's have opposite signs (which occurs only once in the table).
- Column (8) gives the expected frequencies which have been obtained by multiplying the relative frequencies obtained in Col. (7) by the number of observations, *i.e.*, 100 in this case.

Example 22:

The following table gives the distribution of height of first year students of a college :

Height of inches 61 62 63 64 65 66 67 68 69 70 71 72 73 74

Frequency 2 10 11 38 57 93 106 126 1 87 75 23 9 49

Test the normality of the distribution by comparing the proportion of cases bring between $\overline{X} \pm 1\sigma$, $\overline{X} \pm 3\sigma$ *for the distribution and for normal curve.*

Solution:

Calculation of $\overline{X}$ and s

X f d fd fd²

612 –6 –12 72

Solution:

Calculation of $\overline{X}$ and σ

X	f	d	fd	fd²
61	2	–6	–12	72
62	20	–5	–50	250
63	11	–4	–44	176
64	38	–3	–114	342
65	57	–2	–114	228
66	93	–1	–93	93
67	156	0	0	0
68	126	+1	+126	126
69	109	+2	+218	436
70	87	+3	+261	783
71	75	+4	+300	1,200
72	23	+5	+115	575
73	9	+6	+54	324
74	4	+5	+28	196
	N = 750		Σfd = 675	Σ fd² = 4,801

$$\overline{X} - A + \frac{\Sigma fd}{N} = 67 + \frac{675}{750} = 67.9$$

$$\sigma = \sqrt{\frac{\Sigma fd^2}{N} - \left(\frac{\Sigma fd}{N}\right)^2} = \sqrt{\frac{4801}{750} - \left(\frac{675}{750}\right)^2}$$

$$= \sqrt{6.4 - .81} = \sqrt{5.59} = 2.36$$

$\overline{X} = 1\sigma = 67.9 \pm 2.36 = (65.54 \text{ and } 70.26)$

Number of students having height in the range $\overline{X} + 2\sigma$

$= 11 + 38 + 57 + 93 + 106 + 126 + 109 + 87 + 75 + 23 = 725$

$$\text{Proportion} = \frac{725}{750} = 0.967 \text{ or } = 96.7\%$$

$\overline{X} \pm 3\sigma = 67.9 \pm 7.1 = 60.8$ and 75

Number of students having height in the range $\overline{X} \pm 3\sigma = 750$ and hence 100%. In a normal distribution the proportion lying between these limits is 68%, 95% and 100% respectively. Hence the given distribution is approximately normal.

The Normal Distribution as an approximation to describe probability distributions. The normal distribution can be used as an approximation to describe probability distributions such as binomial, the Poisson or the hypergeometric. When such approximation is done it is desirable to make the *Correction for Continuity* adjustment. By revising the approximation formulas so as to include the continuity correction it is also possible to obtain, probability approximations for individual values of the random variable– which otherwise could not be estimated (for details refer to *Business Statistics* by Berenson A Levine).

Example 23:

If the probability of a defective bolt is 0.1, find (a) the mean, and (b) the standard deviation of defective bolts in a total of 900. Also calculate skewness and kurtosis.

Solution:

(a) Mean = np = 900 (0.1) = 90, *i.e.*, we can expect 90 bolts to be defective

(b) Variance = npq = 900 (0.1) (0.9) = 81

Hence the standard deviation = $\sqrt{81} = 9$

$$\text{Skewness} = \frac{q-p}{\sqrt{npq}} = \frac{0.9-0.1}{9} = 0.89$$

$$\text{Kurtosis} = 3 + \frac{1.6pq}{npq} = 3 + \frac{1-6\,(0.1)\,(0.9)}{81} = 3 + 0.006 = 3.006.$$

Since the calculated value is little more than 3, the distribution is slightly *Leptokurtic.*

Example 24:

The following table shows the number of customers returning the products in a marketing territory. The data is for 100 stores:

No. of returns :	*0*	*1*	*2*	*3*	*4*	*5*	*6*
No. of stores :	*4*	*14*	*23*	*23*	*18*	*9*	*9*

Fit a Poisson distribution.

Solution:

Fitting of Poisson Distribution

X	f	fX
0	4	0
1	14	14
2	23	46
3	23	69
4	18	72
5	9	45
6	9	54
	N = 100	Σ fx = 300

$$\overline{X} = \frac{\Sigma fX}{N} = \frac{300}{100} = 3$$

Hence the mean of Poisson distribution, *i.e.*, m = 3.

$P(0) = e^{-m} = 2.7183^{-3}$

$= \text{Rec. [AL (log } 2.7183 \times 3)]$

$= \text{Rec. [AL } (.4343 \times 3)]$

$= \text{Rec. [AL } (1.3029)] = \text{Rec. } 20.08 = 0.0498$

$NP(0) = (P_0) \times N = .0498$

$NP_{(1)} = NP(0) \times m = 5 \times 3 = 15.0$

$$NP_{(2)} = NP_{(1)} \times \frac{m}{2} = 15 \times \frac{3}{2} = 22.5$$

$$NP_{(4)} = NP_{(3)} \times \frac{m}{4} = 22.5 \times \frac{3}{4} = 16.9$$

$$NP_{(5)} = NP_{(4)} \times \frac{m}{5} = 16.9 \times \frac{3}{5} = 10.1$$

$$NP_{(6)} = NP_{(5)} \times \frac{m}{6} = 10.1 \times \frac{3}{6} = 5.1.$$

Example 25:

Bring out the fallacy, if any, in the following statements :

(a) The mean of a binomial distribution is 15 and its standard deviation 5.

(b) Find the binomial distribution whose mean is 6 and variance 4.

Solution:

(a) Mean of binomial distribution is np and standard deviation $\sqrt{npq}$

Given $np = 15$ and $\sqrt{npq} = 5$

Since $\sqrt{npq} = 5$, $npq = 25$

$$q = \frac{25}{np} = \frac{25}{15} = 1.67.$$

We know that p + 1 = 1. Hence q can never exceed 1. But in this question q is greater than 1 (1.67) which is not possible. The statement is, therefore, wrong.

(b) We are given :

$np = 6$ ...(i)

$npq = 4$ ...(ii)

Dividing (ii) by (i), we have

$$q = \frac{4}{3} = \frac{2}{3} \text{ and } p = 1 - \frac{2}{3} = \frac{1}{3}$$

Substituting the value of p

$$n \times \frac{1}{3} = 6$$

$n = 6 ' 3 = 18$

Hence the required binomial distribution is

$$(q + p)^n = \left(\frac{2}{3} + \frac{1}{3}\right)^{18}$$

Example 26:

The incidence of occupational disease in an industry is such that the workmen have a 20% chance of suffering from it. What is the probability that but of six workmen, 4 or more will contact the disease?

Solution:

Let the random variable x denote the number of workers suffering from the disease. The possible values of x are 0, 1, 2, ..., 6.

P, *i.e.*, Probability of a worker suffering from a disease $= \frac{20}{100} = \frac{1}{5}$

$$q = 1 - p = 1 - \frac{1}{5} = \frac{4}{5}$$

Applying binomial distribution

$$P(r) = NC_r q^{n-r} p^r$$

$$N = 6,\ q = \frac{4}{5},\ p = \frac{1}{5}$$

Probability that 4 or more workers would contact the disease is :

$$P(x \le 4) = P(x = 4) + P(x = 5) + P(x = 6)$$

$$= {}^6C_4\left(\frac{1}{5}\right)^4 + {}^6C_5\left(\frac{1}{5}\right)^5\left(\frac{4}{5}\right) + {}^6C_6\left(\frac{1}{5}\right)^6$$

$$= \frac{15 \times 16}{15625} + \frac{6 \times 4}{15625} + \frac{1}{15625} = \frac{265}{15625} = 0.0169$$

Example 27:

Out of 800 families with 4 children each, what percentage would be expected to have (a) 2 boys and 2 girls, (b) at least one boy, (c) no girls, and (d) at the most 2 girls. Assume equal probabilities for boys and girls.

Solution:

(i) Probability of getting a boy $= \frac{1}{2}$ i.e., $p = \frac{1}{2}$

Probability of getting a girl $= \frac{1}{2}$ i.e., $q = \frac{1}{2}$

Probability of getting 2 boys and 2 girls

$$= {}^nC_r q^{n-r} p^r$$

$$= {}^4C_2\left(\frac{1}{2}\right)^2\left(\frac{1}{2}\right)^2$$

$$= \frac{4 \times 3}{2 \times 1} \times \frac{1}{16} = \frac{3}{8}.$$

Percentage of families expected to have two boys and two girls

$$= \frac{3}{8} \times 100 = 37.5\%.$$

(ii) Probability of getting at least one boy means the sum of the probabilities of getting one boy and 3 girls, 2 boys and 2 girls, 3 boys and 1 girl and 4 boys and no girl.

Probability (1 boy and 3 girls) = $^nC_r\, q^{n-r} p^r$

$$= {}^4C_1\left(\frac{1}{2}\right)^2\left(\frac{1}{2}\right)^2 = \frac{4}{16} = \frac{1}{4}$$

probability (2 boys and 2 girls) = $\frac{3}{8}$

probability (3 boys and 1 girl) = ${}^4C_3\left(\frac{1}{2}\right)^1\left(\frac{1}{2}\right)^2 = \frac{1}{4}$

probability (4 boys and no girl) = $\left(\frac{1}{2}\right)^4 = \frac{1}{16}$

$\therefore$ Probability of getting at least one boy = $\frac{1}{4}+\frac{3}{8}+\frac{1}{16}=\frac{15}{16}$

Percentage of families expected to have at least one boy

$$= \frac{15}{16} \times 100 = 93.75\%$$

(iii) Percentage of families expected to have no girls, *i.e.*, all boys

$$= \frac{1}{16} \times 100 = 6.25$$

(iv) Probability of getting at most 2 girls = $= \frac{1}{16}+\frac{1}{4}+\frac{3}{8}=\frac{11}{16}$

$\therefore$ Percentage of families expected to have at most 2 girls

$$= \frac{11}{16} \times 100 = 68.75.$$

Example 28:

A cigrarette company wants to promote the sales of X's cigarettes (Brand) with special advertising campaign. Fifty out of every thousand cigarettes are rolled up in gold foil and randomly mixed with the regular (special king-sized, mentholated) cigarettes. The company offers to trade a new package of cigarettes for each gold cigarette a smoker finds in a package of Brand X. What is the probability that buyers of Brans X will find X = 0, 1, 2, 3,.. gold cigarettes in a single package of 10?

Solution:

This can be considered a Poisson experiment in which there are s = 10 trials in a pack and the probability of finding a golden cigarette (a success) is p = 50/100 = 0.05. The expected number of golden cigarettes per pack is, therefore, m = sp = 10 (0.05). Using the table of the Poisson distribution, we obtain

Number of golden cigarettes per pack	Probability
0	0.6065
1	0.3033
2	0.0758
3	0.0126
4	0.0016

We find that 60.65 per cent of the packages contain no golden cigarettes, 30.33 per cent contain one golden cigarette, 7.58 per cent contain 2 golden cigarettes, 1.26 per cent contain 3 golden cigarettes and 0.16 per cent contain 4 golden cigarettes.

Example 29:

You are incharge of rationing in a State affected by food shortage. The following reports were received from investigators:

Daily calories of food available per adult during current period

Area	*Mean*	*S.D.*
A	*2,000*	*350*
B	*1,750*	*100*

The estimated requirement of an adult is taken at 2,500 calories daily and the absolute minimum of 1,000. Comment on the reported figures and determines which area in poor opinion needs more urgent attention.

Solution:

Area A	Area B
Mean ± 3σ	Mean ± 3σ
2,000 ± (3 × 350)	1,750 ± (3 × 100)
Between 950 and 3,050 calories	Between 1,450 and 2,050 calories

Since the estimated requirement is minimum of 1,000 calories, areas A needs more urgent attention because there are people here who are getting less than 1,000 calories.

Example 30:

Eight coins are thrown simultaneously. Find the chance of obtaining.

(i) at least 6 heads,

(ii) no heads, and

(iii) all heads.

Solution:

(i) In tossing 8 coins simultaneously, the probability of getting at least six heads will be given by the sum of separate probability of getting 6 heads, no heads and 8 heads.

$$n = 8,\ r\ 6,\ 7,\ 8,\ p = \frac{1}{2}, q = \frac{1}{2}$$

$$p\ (r = 6, 7, 8) = {}^8C_8\left(\frac{1}{2}\right)^5\left(\frac{1}{2}\right)^2 + {}^2C_1\left(\frac{1}{2}\right)^7\left(\frac{1}{2}\right) + {}^3C_3\left(\frac{1}{2}\right)^3\left(\frac{1}{2}\right)^3$$

$$= \left(28 \times \frac{1}{256}\right) + \left(8 \times \frac{1}{256}\right) + \left(1 \times \frac{1}{256}\right) = \frac{37}{256}$$

(ii) Probability of getting no head is given by

$${}^2C_0\left(\frac{1}{2}\right)^0\left(\frac{1}{2}\right) = 1 \times \frac{1}{256} = \frac{1}{256}$$

(iii) Probability of getting all heads is given by

$${}^3C_3\left(\frac{1}{2}\right)^2\left(\frac{1}{2}\right)^3 = 1 \times \frac{1}{256} = \frac{1}{256}.$$

Example 31:

The normal rate of infection of a certain disease in animals is known to be 25%. In an experiment with 6 animals injected with a new vaccine it was observed that none of the animals caught infection. Calculate the probability of the observed result.

Solution:

Let p denote infection of the disease

$$p = \frac{25}{100} = \frac{1}{4} \text{ and } q = 1 - \frac{1}{4} = \frac{3}{4}$$

Out of 6 animals probability of 0, 1, 2, 3, 4, 5 and 6 animals catching infection will be given by 1st, 2nd, 3rd terms, etc., in the expansion of $\left(\frac{3}{4} + \frac{1}{4}\right)^6$

$\therefore$ The probability of none of the animals being infected $= \left(\frac{3}{4}\right)^6 = \frac{729}{4096}$.

Example 32:

It is given that 3% of electric bulbs manufactured by a company are defective. Using the Poisson approximation, find the probability that a sample of 100 bulbs will contain (i) no defective, (ii) exactly one defective.

Solution:

Number of defective bulbs in a sample of 100 is 3 (since 3% bulb manufactured are defective).

Hence $m = 3$

Probability of no defective bulb in a sample of 100 is given by

$$P(0) = e^{-m} = e^{-3} = 0.05 \text{ (from the table)}$$

$$P(1) = P(0) \times m = 0.05 \times 3 = 0.15.$$

Example 33:

A multiple-choice test consists of 8 questions with 3 answers to each question (of which only one is correct). A student answers each question by rolling a balanced dice and checking the first answer if he gets 1 or 2, the second answer if he gets 3 or 4 and the third answer if the gets 5 or 6. To get a distinction, the student must secure at least 75% correct answers. If there is no negative marking, what is the probability that the student secures a distinction?

Solution:

Probability of getting a correct answer to a question.

i.e., $p = \frac{1}{3}$ and $q = \frac{2}{3}$

By Binomial distribution the probability of getting correctly r answers in 8 questions test is given by

$$p(r) = {}^8C_1 p^r q^{B-r}; \; r = 0, 1, 2, ..., 8$$

The required probability 'p' of securing a distinction (*i.e.*, of getting correct answers of at least 6 of the 8 questions) is given by:

$$P = P(6) + P(7) + P(8)$$

$$= {}^8C_6\left(\frac{1}{2}\right)^6\left(\frac{2}{3}\right)^6 + {}^8C_7\left(\frac{1}{3}\right)^7\left(\frac{2}{3}\right) + {}^8C_8\left(\frac{1}{3}\right)^8$$

$$= \frac{1}{36}\left[\left(27 \times \frac{4}{9}\right) + \left(8 \times \frac{1}{3} \times \frac{2}{3}\right) + \frac{1}{9}\right]$$

$$= \frac{1}{729}\ (12 + .178 + 0.111)$$

$$= \frac{13.889}{729} = 0.019.$$

Example 34:

Is there any inconsistency in the statement, the mean of binomial distribution is 20 and its standard deviation 4? If no inconsistency is found what shall be the values of p, q and n?

Solution:

The mean of the binomial distribution is given by np and standard deviation by $\sqrt{npq}$

Here $\quad np = 20$ and $\sqrt{npq} = 4,$

Since $\sqrt{npq} = 4,\ npq = 16$

Putting the value of np

$$20q = 16 \text{ or } q = \frac{16}{20} = 0.8$$

$$q = 8,\ p = 1 - .8 = 0.2$$

Since $\quad npq = 16$

By putting the value of p and q, we can find n,

$$n \times 2 \times .8 = 16\ n = \frac{16}{.16} = 100.$$

Example 35:

Certain mass-produced articles of which 0.5 per cent are defective, are packed in cartons each containing 130 articles. What proportion of cartons are free from defective articles, and what proportion contain 2 or more defectives. (Given $e^{22} = 0.6065$*).*

Solution:

$$P(r) = \frac{e^{-m}m^r}{r!}, \; r = 0, 1, 2, ...$$

$P(0) = e^{-0.5} = 0.6065$ (given)

$P(1) = P(0) \times m = 0.6065 \times 0.5 = 0.30325.$

The probability of a carton with 2 or more defectives

$= 1 - 0.6065 - 0.30325 = 0.09025$

Therefore the proportion of cartons free from defective articles is 80.65% or 61% and with 2 or more defectives is 9% (approx).

Example 36:

The following table gives frequencies of occurrence of a variate x between certain limits :

Variate (X)	*Frequency (f)*
Less than 40	*30*
40 or more but less than 50	*33*
50 and more	*37*
	100

The distribution is exactly normal. Find the average and standard deviation of X.

Solutlon:

Area between the ordinate 40 and $\overline{X}$ = (50 – 30)% = 20%

Area between the ordinate 50 and $\overline{X}$ = (50 – 37) = 13%

For the proportion of area 0.20. $\frac{\left(X-\overline{X}\right)}{\sigma} = 0.5244$

For the proportion of area 0.13. $\frac{X-\overline{X}}{\sigma} = 0.3318$

$$\frac{40-\overline{X}}{\sigma} = -0.5244$$

and $\quad 50-\frac{\overline{X}}{\sigma} = 0.3318$

Substracting (i) from (ii) $10 = 9.8562\sigma$ or $\sigma = 11.68$

Putting the value of s in equation (i)

$$\overline{X} - 40 = .5244 \times 11.68 \text{ or } \overline{X} = 40 + 6.12 = 46.12.$$

Example 37:

The average monthly sales of 5000 firms are normally distributed. Its mean and standard deviation are Rs. 36000 and Rs. 10000 respectively. Find

(i) the number of firms the sales of which are over Rs. 40000.

(ii) the percentage of firms the sales of which will be between Rs. 38500 and 41000.

(iii) the number of firms the sales of which will be between Rs. 30000 and Rs. 40000. The relevant extract at the area table (under the normal curve) is given below :

Z	*0.25*	*0.40*	*0.5*	*0.6*
Area	*0.0987*	*0.1554*	*0.1915*	*0.2257*

Solution:

(i) The number of firms whose sales are over Rs. 40000.

$N = 5000, \overline{X} = 36000, \sigma = 10000, X = 40000$

$$Z = \frac{X - \overline{X}}{\sigma} = \frac{40000 - 36000}{10000} = 0.4$$

Area between Z = 0 and Z = 0.4 will be 0.1554.

Area above Z = 0.4 = 0.5 − 0.1554 = 0.3446

No of firms whose sales are over Rs. 40000

= 5000 × .3446 = 1723

(ii) Percentage of firms whose sale is between Rs. 38500 and Rs. 41000

$$Z = \frac{38500 - 36000}{10000} = \frac{2500}{10000} = 0.25$$

Area = 0.987

$$\text{Again } Z = \frac{41000 - 36000}{10000} = 0.5$$

Area = 0.1915

∴ Area between 38500 & 41000 = .1915 − .0987

= 0.0928

Hence percentage = 9.28

(iii) Number of firms where sales is between Rs. 30,000 and Rs. 40,000

$$z = \frac{30{,}000 - 36{,}000}{10{,}000} = -0.6$$

Area = 0.2257

$$\text{Again } Z = \frac{40{,}000 - 36{,}000}{10{,}000} = 0.4$$

Area = 0.1554

Area between 30,000 and 40,000

= 0.2257 × 0.1554 = 0.3811

∴ No of firms = 0.3811 × 5,000 = 1905.5 or 1906.

Example 38:

When the first proof of 200 pages of an encyclopaedia of 5,000 pages was read, the distribution of printing mistakes was found to be as shown in the first and second columns of the table below. Fit a Poisson distribution to the frequency distribution of printing mistakes. Estimate the total cost of correcting the whole encyclopaedia by using the information given in the first and third columns of the table below :

No. of misprints on a page	***Frequency***	***Cost of defection and correction per page (Rs.)***
0	*113*	*1.00*
1	*62*	*2.50*
2	*20*	*1.50*
3	*3*	*3.00*
4	*1*	*3.50*
5	*1*	*4.00*

Solution:

Fitting of Poisson Distribution

No. of misprints on a page	**Frequency**	**fX**
0	113	0
1	62	62
2	20	40
3	3	9
4	1	4
5	1	5
	N = 200	Σ fX = 120

$$\overline{X} = \frac{\Sigma fX}{N} = \frac{120}{200} = 0.6$$

This is the mean of the Poisson distribution, *i.e.*, m.

$P_3 = e^{-m} = 2.7183^{-6}$

$= \text{Rec. [AL (log } 2.7183 \times .6)] = \text{Rec. [Al } (.6 \times .4343)]$

$= \text{Rec. [AL } .2605] = \text{Rec. } 1.822 = 0.5488$

$N(P_0) = 200 \times .5488 = 109.76$

$N(P_1) = N(P_0) \times m = 109.76 \times .6 = 65.856$

$$N(P_2) = N(P_1) \times \frac{m}{2} = 65.856 \times \frac{.6}{2} = 19.756$$

$$N(P_3) = N(P_2) \times \frac{m}{3} = 19.757 \times \frac{.6}{3} = 3.591$$

$$N(P_4) = N(P_3) \times \frac{m}{4} = -3.951 \times \frac{.6}{4} = 0.5927$$

$$N(P_5) = N(P_4) \times \frac{m}{5} = 0.5927 \times \frac{.6}{5} = 0.0711$$

The frequencies of the Poisson distribution are :

X	0	1	2	3	4	5
f	109.8	65.9	19.8	4.0	0.6	0.07

The total cost of correcting the first proof of the whole encyclopaedia will be given by:

No. of misprints on a page	Rate per page X	No. of pages f	fX
0	1.00	109.80	109.80
1	1.50	65.90	98.85
2	2.50	19.80	49.50
3	3.00	4.00	12.00
4	3.50	.60	2.10
5	4.00	0.07	0.28
			Σ fX = 272.53

Hence the total cost shall be Rs. 272.53

Example 39:

The distribution of typing mistakes committed by a typist is given below: Assuming a Poisson mode, find out the expected frequencies :

No. of mistakes per page :	*0*	*1*	*2*	*3*	*4*	*5*
No. of pages :	*142*	*156*	*69*	*27*	*5*	*1*

Solution:

Calculation of Expected Frequencies by Assuming Poisson Model

No. of Mistakes X	per page f	Frequency fX
0	142	0
1	156	156
2	69	138
3	27	18
4	5	20
5	1	5
	N = 400	Σ fX = 400

$$\overline{X} = \frac{\Sigma fX}{N} = \frac{400}{400} = 1$$

This is the mean of Poisson distribution, *i.e.*, m

$$P_{(0)} = e^{-m} = 2.7183^{-1} = \frac{1}{2.7183} = 0.3679$$

$$N(P_0) - 400 \times .3679 - 147.16$$

$$N(P_1) = N(P_3) \times m = 147.16 \times 1 = 147.16$$

$$N(P_2) = N(P_1) \times \frac{m}{2} = 1.47.16 \times \frac{1}{2} = 73.58$$

$$N(P_3) = N(P_2) \times \frac{m}{3} = 73.58 \times \frac{1}{3} = 24.53$$

$$N(P_4) = N(P_3) \times \frac{m}{4} = 24.53 \times \frac{1}{4} = 6.13$$

$$N(P_5) = N(P_4) \times \frac{m}{5} = 6.13 \times \frac{1}{5} = 1.23.$$

Thus the expected frequencies as per the Poisson model approximated to the whole number are :

No. of mistakes per Page :	0	1	2	3	4	5
No. of pages:	147	147	74	25	6	1

Example 40:

Find the probability that the value of an item drawn at random from a normal distribution with mean 20 and standard deviation 10 will be between:

(a) 10 and 15 (b) – 5 and 10, and (ii) 15 and 25.

The relevant extract of the Area Table (under the normal curve is given below) :

0.5	1.0	1.5	2.0	2.5
0.1915	0.3413	0.4332	0.4772	0.4938

Solution:

We are given $\overline{X} = 20$ and $\sigma = 10$

(i) For X = 10, $Z_1 = \frac{X-\overline{X}}{\sigma} = \frac{10-20}{10} = -1$

For X = 15, $Z_2 = \frac{X-\overline{X}}{\sigma} = \frac{15-20}{10} = -0.5$

$\therefore P = (10Z \times < 15) = 0.3413 - 0.1915 = 0.1498$

(ii) For X = – 5, $Z_1 = \frac{X-\overline{X}}{\sigma} = \frac{-5-20}{10} = -2.5$

For X = 10, $Z_2 = \frac{X-\overline{X}}{\sigma} = \frac{10-20}{10} = -1$

$\therefore P(-5 < x < 10) = 0.4938 - 0.3413 = 0.1525$

(iii) For X = 15, $Z_1 = \frac{X-\overline{X}}{\sigma} = \frac{15-20}{10} = -0.5$

For X = 25, $Z_2 = \frac{X-\overline{X}}{\sigma} = \frac{25-20}{10} = +0.5$

$\therefore P(15 < x < 25) = 2x - 1915 = 0.3830.$

Example 41:

The income distribution of workers in a certain factory was found to be normal with mean of Rs. 500 and standard deviation equal to Rs. 50. There were 228 persons getting above Rs. 600. How many persons were there in all? (Area under the standard normal curve between heights at 0 and 2 is 0.4772).

Solution:

The probability of the persons getting above Rs. 600, *i.e.*,

$$P\ (X > 600) = 1 - P\ (X \leq 600)$$

$$= 1 - P\left(z \leq \frac{600 - 500}{50}\right)$$

$$= P\ (Z \leq 2)$$

$$= 1 - \{P\ (Z \leq 0) + P(0 \leq Z \leq 2)\}$$

$$= 1 - \{0.5000 + 0.4772\}$$

$$= 1 - 0.9772 = .0228$$

Since there are 228 persons getting salary above Rs. 600, it means that the total number of persons is:

$$228/0.0228 = 10{,}000.$$

Example 42:

Among 10,000 random digits, in how many cases do we expect that the digit 3 appears at most 950 times. (The area under standard normal curve for z = 1.667 is 0.4524 approximately).

Solution:

$n = 100$

Probability of digit 2 appearing = 1/10

Hence $p = 1/10,\ q = 1 - 1/10 = 9/10$

$$np = 1000\ '\ 1/10 = 1000$$

$$npq = 10000 \times \frac{1}{10} \times \frac{9}{19} = 900$$

Let n be the number of times the digit 3 appears among 10,000 random digits. Then n follows binomial distribution with mean 1000 and variance 900.

$$\therefore P\ (n \leq 950) = p\left(\frac{Z \leq 950 - 1000}{\sqrt{900}}\right)$$

$$= p(Z \leq -1.667)$$

$$= 0.5000 - 0.4525 = 0.475$$

No. of cases = $10000 \times 0.0475 = 475$

Hence in 475 cases we may expect the digit to appear at most 950 times.

(b) Point out the fallacy if any in the following statement "The mean of a binomial distribution is 10 and its standard deviation is 4".

Solution:

The mean of binomial distribution is np and standard deviation $\sqrt{npq}$.

We are given $np = 10, \sqrt{npq} = 4$

$\sqrt{npq} = 4$ or npq 16

Putting the value of np,

$$10q = 16 \text{ or } q = \frac{16}{10} = 1.6$$

Since the value of q is exceeding one, there is some inconsistency in the given statement.

Example 43:

In a town 10 accidents took place in a span of 50 days. Assuming that the number of accidents per day follows the Poisson distribution, find the probability that there will be three or more accidents in a day.

Solution :

The average number of accidents per day

$$= \frac{10}{50} = 0.2$$

P(3 or more accidents) = 1 – P(2 or less accidents)

$$= 1 - [p(0) + P(1) + P(2) + P(2)]$$

$$= 1 - \left[e^{-2} + e^{-2} \times 0.2 + \frac{e^{-2} \times 0.2 \times .2}{2}\right]$$

$$= - e^{-2} [1 + .2 + 0.02]$$

$$= - e^{-2} \times 1.22$$

$$= .8187 \times 1.22$$

$$= 1 - .999 = 0.001$$ (From table of e^{-m}).

Example 44 :

The customer accounts at a certain departmental store have an average balance of Rs. 480 and a standard deviation of Rs. 160. Assuming that the account balances are normally distributed.

(i) What proportion of the accounts is over Rs. 600?

(ii) What proportion of the accounts is between Rs. 400 and Rs. 600?

(iii) What proportion of the accounts is between Rs. 240 and Rs. 360?

Solution:

Let the random X denote the balance of the customer accounts. X is normally distributed with $\overline{X}$ = 480 and σ = 160. The standard normal variable Z is :

$$Z = \frac{X - \overline{X}}{\sigma} = \frac{X - 480}{100}$$

(i) when X = 600

$$Z = \frac{600 - 480}{160} = 0.75$$

P(X > 600) = P(Z > 0.75)

= Area of the right of Z = 0.75

= 0.5000 – 0.2734 = 0.2266

Hence 22.6% of the accounts have a balance in excess of Rs. 600.

(ii) Probability that the accounts lie between Rs. 400 and Rs. 600 is given by:

P(400 ≤ X ≤ 600)

when X = 400

$$Z = \frac{400 - 480}{160} = -05$$

when X = 600

$$Z = \frac{400 - 480}{160} = 0.75$$

P(400 < x < 600)

= P(1 – 0.5 < Z < 0.75)

Area between Z = – 0.5 to Z = 0.75

= Area between Z = – 0.5 to Z = 0) + (Area between

= Z = 0 to Z = 0.75

= P (– 0.5 ≤ Z ≤ 0) + P (0 ≤ Z ≤ 0.75)

= 0.1915 + 0.2734 = 0.4649

Hence 46.69% of the accounts have an average balance between Rs. 400 and Rs. 600.

(C) We want P(240 ≤ X ≤ 360)

(Hint when X = 240 find Z, when X = 360 find Z. Calculate area between Z = – 0.75 to Z = – 1.5 Ans. 15.98 per cent.)

Example 45:

Calculate the frequencies of the normal distribution which was the same mean, standard deviation and total frequency as the distribution given below or the intervals 60 – 65 – 70, etc.

X	*60 –*	*65 –*	*70 –*	*75 –*	*80 –*	*85 –*	*90 –*	*95 –*
f	*3*	*21*	*150*	*335*	*325*	*135*	*26*	*4*

Solution:

For finding frequencies of the normal distribution, first we will find out mean and standard deviation of the given distribution :

Calculation of mean and Standard Deviation

Variable	**m.**	**f**	**(m – 77.5)/5** **d**	**fd**	**fd²**
60 – 65	62.5	3	–3	–9	27
65 – 70	67.5	21	–2	–42	84
70 – 75	72.5	150	–1	–150	150
75 – 80	77.5	335	0	0	0
80 – 85	82.5	326	+1	+326	326
85 – 90	87.5	135	+2	+270	540
90 – 95	92.5	26	+3	+78	234
95 – 100	97.5	4	+4	+16	64
		N = 1,000		Σ fd = 489	Σ fd² = 1425

$$\overline{X} = A + \frac{\Sigma\, fd}{N} \times i = 77.5 + \frac{489}{1000} \times 5 = 79.945 \text{ or } 79.95$$

$$\sigma = \sqrt{\frac{\Sigma\, fd^2}{N} - \left(\frac{\Sigma\, fd}{N}\right)^2} \times i = \sqrt{\frac{1425}{1000} - \left(\frac{489}{1000}\right)^2} \times 5$$

$$= \sqrt{1.425 - .239} \times 5 = 1.089 \times 5 = 5.45$$

Fitting of Normal Curve by Method of Areas

Variable	Lower Limits X	f	(X – 79.95 x	$\frac{x}{\sigma}$	Area Under Normal Curve from 0 to Z	Area for each class	Expected frequencies
60–65	60	3	– 19.95	–3.66	0.4999	0.0030	.0030 × 1000 = 3.0
65–70	65	21	– 14.95	–2.74	0.4969	0.0305	.0305 × 1000 = 30.5
70–75	70	150	–9.95	–1.83	0.4664	0.1480	.1480 × 1000 = 148.0
75–80	75	335	0.05	– 0.91	0.3186	0.3146	.3146 × 1000 = 314.6
80–85	80	326	+ 5.05	– 0.01	0.0040	0.3278	.3278 ′ 1000 = 327.8
85–90	85	135	+ 10.05	+ 0.93	0.3238	0.1433	.1433 × 1000 = 143.3
90 – 95	90	26	+ 15.05	+ 1.84	0.4671	0.0300	.0300 × 1000 = 30.0
95 – 100	95	4	+ 20.05	+ 2.76	0.4971	0.0028	.0028 × 1000 = 2.8
	100			+ 3.68	0.4999		
	N = 100					Total 1000	

Thus the frequencies of the normal distribution are :

X	60–65	65–70	70–75	75–80	80–85	85–90	90–95	95–100
f	3	30.5	148.5	314.5	327.8	143.3	30.0	2.8

Example 46 :

Fit a Poisson distribution to the following data and calculate theoretical frequencies :

Deaths	*0*	*1*	*2*	*3*	*4*
Frequency	*122*	*60*	*15*	*2*	*1*

Solution:

Fitting of Poisson Distribution

Deaths X	frequency f	fX
0	122	0
1	60	60
2	15	30
3	2	6
4	1	4
N = 200	Σ fX = 100	

$$m = \frac{\Sigma fX}{N} = \frac{100}{200} = 0.5$$

$$P_{(0)} = e^{-m} = e^{-0.5} = .6065$$

$$NP_{(0)} = N \times P_{(0)} = 200 \times .6065 = 212.30 \text{ or } 121$$

$$NP_{(1)} = NP_{(0)} \times m = 121.3 \times .5 = 65.65 \text{ or } 61$$

$$NP_{(2)} = NP_{(1)} \times \frac{m}{2} = 60.65 \times \frac{.5}{2} = 15.16 \text{ or } 15$$

$$NP_{(3)} = NP_{(2)} \times \frac{m}{3} = 15.16 \times \frac{.5}{3} = 2.53 \text{ or } 3$$

$$NP_{(4)} = NP_{(3)} \times \frac{m}{4} = 2.53 \times \frac{.5}{4} = .032 \text{ or } 0.$$

Example 47:

In a manufacturing organisation, the distribution of wages was perfectly normal and the number of workers employed in the organisation was 5000. The mean wages of the workers were calculated at Rs. 800 P.M. and the standard deviation was worked out to Rs. 200. On the basis of the information estimate :

(i) the number of workers getting salary between Rs. 700 and Rs. 900.

(ii) Percentage of workers getting salary above Rs. 1000.

(iii) Percentage of workers getting salary below Rs. 600.

Solution:

(i) $$Z_1 = \frac{X - \overline{X}}{\sigma} = \frac{700 - 800}{200} = -0.5$$

$$Z_2 = \frac{X - \overline{X}}{\sigma} = \frac{900 - 800}{200} = 0.5$$

Area at $Z = -0.5 = 0.1915$

Area at $Z = +0.5 = 0.1915$

Total $= 0.3830$

Number of workers getting wages between Rs. 700 and 900 = .383 × 5000 = 1915

(ii) Percentage of workers getting salary above Rs. 1000.

$$Z = \frac{1000 - 800}{200} = 1$$

Area to the right of $\overline{X}$ at Z = 1 = 0.5 – 0.3413 = 0.1587

No. of workers = 0.1587 × 5000 = 794

∴ Percentage of workers getting salary more than 1000

$$= \frac{794}{5000} \times 100 = 15.87$$

(iii) Percentage of workers getting below Rs. 600

$$Z = \frac{X - \overline{X}}{\sigma} = \frac{600 - 800}{200} = -1.0$$

Area to the left of $\overline{X}$ at Z = 1 = 0.5000 – 0.3413 = 0.1587

No. of workers = 0.1587 × 5000 = 793.5

Percentage of workers getting salary below 600

$$= \frac{793.5}{5000} \times 100 = 15.87.$$

Example 48:

A aptitude test for selecting officers in a bank was conducted on 1,000 candidates, the average score is 42 and the standard deviation of scores is 24.

Assuming normal distribution for the scores, find :

(a) the number of candidates whose scores exceed 58.

(b) the number of candidates whose scores lie between 30 and 66.

Solution:

(A) Number of candidates whose score exceeds 58.

$$Z = \frac{X - \overline{X}}{\sigma} = \frac{60 - 42}{24} = 0.667$$

Area to the right of 0.667 is (0.5 – 0.2476) = 0.2524

∴ Number of candidates whose score exceeds 60 is

= 100 × 0.2524 = 252.4 or 252

(b) Number of candidates whose score exceeds 58.

$$Z = \frac{30 - 42}{24} = -0.5$$

Standard normal variate corresponding to 66

$$Z = \frac{66 - 42}{24} = 1$$

Area between Z = – 0.5 and Z = 1

= .1915 + 0.3413 = 0.5328

Number of candidates whose score lie between 30 and 66

= 1000 × 0.5328 = 532.8 or 533.

Example 49:

The mean of a binomial distribution is 20 and standard deviation 4, calculate n, p, q.

Solution:

The mean of binomial distribution np and standard deviation $\sqrt{npq}$.

Hence np = 20 and

$\sqrt{npq}$ = 4 or = npq = 16

Since np is 20

20q = 16

or q = 16/20 = 0.8

p = 1 – q = 1 .8 = 0.2

n × .2 × .8 = 16

$$n = \frac{16}{.16} = 100$$

Hence n = 100,

p = .2, q = .8

Example 50:

Of a large group of men 5 per cent are under 60 inches in height and 40 per cent are between 60 and 65 inches. Assuming a normal distribution, find the mean height and standard deviation.

Solution:

The question can easily be understood from the following diagram :

Since the area lying to the left of X = 60 is 0.05 the area between z = 0 and z_2 = (60 – $\overline{x}$)/σ is 0.50 – 0.45. From the table, the value of z_1 corresponding to this area is 1.645. Therefore,

$$\frac{60 - \overline{X}}{\sigma} = -1.645 \qquad ...(i)$$

Also the height of 40% of the men is between 60 and 65 inches, the area between the ordinates at X = 60 and X = 65 is 0.4 or the area to the left of the ordinate at X – 65 is 0.45. The area between z = 0 and z_2 = (65 – $\overline{X}$)/σ is 0.50 – 0.45. The value of z_2 corresponding to this area is 0.13.

$$\frac{65-\overline{X}}{\sigma} = -0.13$$

Solving equations (i) and (ii)

$$60 - \overline{X} = -1.645\sigma$$

$$65 - \overline{X} = 0.130\sigma$$

$$-5 = -1.515\sigma$$

or $\sigma = \frac{5}{1.515} = 3.3$

Substituting the value of s in eq. (ii)

$$65 - \overline{X} = 0.13\ (3.3)$$

$$\overline{X} = -65 - 429$$

or $\overline{X} = 65.429.$

Example 51:

The mean inside diameter of a sample of 500 washers produced by a machine is 5.02 mm and the standard deviation is 0.05 mm. The purpose for which these washers are intended allows a maximum tolerance in the diameter of 4.96 to 5.08 mm. otherwise the washers are considered defective. Determine the percentage of defective washers produced by the machine, assuming the diameters are normally distributed.

Solution:

4.96 in standard units = (4.96 – 5.02)/0.05 = – 1.2

5.08 in standard units = (5.08 – 5.02)/0.05 = 1.2

Proportion of non-defective washers

= (area under the normal curve between Z = 1.2 and Z = 1.2)

= (twice the area between Z = 0 and Z = 1.2)

= 2 (0.38497) = 0.7699 or 77%

Thus the percentage of defective washers = 100 – 77 = 23.

EXERCISES

1. Below are given the number of vacancies of judges occurring in a High Court over a period of 96 years :

No. of Vacancies	:	0	1	2	3	Total
Frequency	:	59	27	9	1	96

Find Poisson Distribution to represent the frequencies of vacancies per car.

2. (a) What is normal distribution? Point out its important properties.

(b) Fit a Binomial distribution to the following data :

X	:	0	1	2	3	4
f	;	28	62	45	10	4

3. A typist takes on an average 8 minutes to type a letter. It is found that she is idle for 20% of the time. (Given : P_t ($Z \leq$ 2,76) = 0.9971)

(i) what would be the number of letters she may be receiving on an average per hour assuming Poisson arrival and exponential servicing?

(ii) What is the time (in minutes) on an average that a letter arrived will have to be on her tray before it is taken by her typing?

(iii) What is the probability that a letter sent to her would be received back fully typed within 5 minutes?

4. (a) If the distribution of incomes of a group of persons be assumed to be normal with mean Rs. 500 and the standard deviation Rs. 50, estimate the proportion of individual with income (i) between Rs. 550 and Rs. 650, (ii) between Rs. 450 and Rs. 75.

(b) The scores made by candidates in a certain test are normally distributed with mean 500 and standard deviation 100. What per cent of candidates receive (i) less than 400, (ii) between 400 and 500?

[(a) (i) 15.735%, (ii) 14.28; (b) (i) 15.86%, (ii) 68.28%]

5. In a super market two girls are attending at the payment counters. If the service time for each customer is exponential with mean of the 4 minutes, and people arrive in Poisson manner at the rate of 10 per hour :

(i) What is the probability of customer having to wait for service?

(ii) What is the expected percentage of idle time for each girl?

6. (a) Given that the number of accidents per day has a Poisson distribution with a mean of 0.4 and that the probability of no

accident on a day is 0.67, what is the probability that there will be only one accident on a particular day?

(b) A binomial distribution has n = 20 and p = 0.3. What are the mean and variance of this distribution?

(c) From a normal distribution with mean 50 a random sample of 20 measurements is selected. How many of these 20 measurements are expected to be less than 50?

7. (a) A manufacturer of electric packs fuses in boxes of 10 each and 2,000 such boxes were sold. The previous experience shows that 5 per cent of the fuses are defective. Using Poisson distribution, find how many boxes will contain (i) no defectives, (ii) more than one defective.

(b) In certain Poisson frequency distribution, the frequency corresponding to successes is half the frequency corresponding to 3 successes. Find its mean and standard deviation.

8. (a) Suppose that a doorway being constructed is to be used by a class of people whose heights are normally distributed with a mean 70" and standard deviation 3". How much high the doorway should be without causing more than 25% of the people to bump their head? If the height of the doorway be fixed at 76", how many persons out of 5,000 are expected to bump their head?

[(a) (i) 635, (ii) 3,274, (a) 72.025", 114 persons]

(b) if 20% of the bolts produced by a machine are defective, determine the probability that out of 4 bolts chosen at random (a) 1, (b) 0, (c) at most 2 bolts will be defective.

[(a) .4096, (b) .4096, (c) .9728]

9. (a) Under what conditions does the Binomial distribution tend to be a normal distribution.

(b) Show that Poisson distribution is the limiting form of Binomial distribution.

(c) Calculate the mean and standard deviation of a binomial distribution with parameter n & p.

10. A manufacturer who produces medicine bottles finds that 0.1% of the bottles is defective. The bottles are packed in boxes each containing 500 bottles. A drug manufacturer buys 100 boxes from the producer of bottles. Using Poisson distribution, find how many boxes will contain :

(i) No defectives

(ii) At least two defectives.

11. The daily wages of 1000 workers are normally distributed around mean of Rs. 70 and with a standard deviation of Rs. 5. Estimate the number of workers whose daily wages will be :

    (i) Between Rs. 70 and Rs. 72;

    (ii) Between Rs. 69 and Rs. 72;

    (iii) More than Rs. 75;

    (iv) Less than Rs. 63

    Also estimate the lowest wages of the 100 highest paid workers.

12. (a) If in a random variable X, follows Poisson distribution such that P (X = 1) = P (X = 2), find (i) the mean of the distribution P (X = 0) (iii) P (X>2).

    (b) The diameter of shafts produced in a factory confirms to normal distribution. 31% of the shafts have diameter less than 45 mm. and 8% have more than 64 mm. Find the mean and standard deviation of the diameter of the shafts produced by the factory.

13. (a) When can Poisson distribution be a reasonable approximation of the binomial?

    (b) "Everybody believes in the normal law of errors : the experimentalist, because he thinks it is a mathematical theorem; the mathematician, because he thinks it is an experimental fact". Discuss this statement and give the chief properties of the Normal distribution.

14. (a) Under what conditions can observed empirical frequency distribution be approximated to Binomial distribution?

    (b) How does a normal distribution differ from a binomial distribution? Mention the properties of a normal distribution. How are they useful in random sampling investigations?

15. (a) Distinguish between observed frequency and theoretical frequency distributions. Why theoretical distribution is considered so essential for solving various business problems.

    (b) What are the main features of Normal probability distribution? Explain the importance of Normal Distribution in statistical theory.

16. (a) What is Bernoulli distribution? What are assumptions to be made while studying it?

(b) Explain the meaning of Benoulli process pointing out its main characteristics.

(c) Explain the circumstances when the normal probability distribution is used in business.

17. (a) What are the characteristics of normal distribution? Describe briefly the importance of normal distribution in statistical analysis.

(b) Define normal distribution. Point out its chief characteristics.

(c) What is normal distribution? Highlight its important properties.

18. The mean yield per plot of a crop is 17 kg. and standard deviation is 3 kg. If distribution of yield per plot is normal, find the percentage of plots giving yields :

(i) between 15.5 kg. and 20 kg., and

(ii) more than 20 kg.

Area between z = 0 and z = 0.5 is 0.1915 and area between z = 1 is 0.3413, where z is standard normal variate.

19. The customer accounts of a certain departmental store have an average balance of Rs. 120 and a standard deviation Rs. 40. Assuming the account balances are normally distributed :

(a) What proportion of the accounts is over Rs. 150?

(b) What proportion of the accounts is over Rs. 150?

(c) What proportion of the accounts is between Rs. 60 and Rs. 90? Given :

Z :	0.50	0.75	1.00	1.50
Area :	0.1915	0.2734	0.3413	0.3531

20. Which of the following statements is True or False :

(a) The mean of binomial distribution is up and its standard deviation npq.

(b) The mean of binomial distribution is 20 and its standard deviation 5.

(c) The normal distribution was first discovered by De Moivre in 1733.

(d) The mean plus or minus 2.85 standard deviation includes 96 per cent of the items in normal distribution.

(e) $(q + p)^5 = q^5 + 5q^4p + 10q^3p^2 + 20q^2p^3 + 5qp^4 + p5$.

(f) As p increase for a fixed n, the binomial distribution shifts to the left.

(g) In a Poisson distribution as n increases, the distribution shifts to the right.

(i) A normal curve is completely defined by the mean and the standard deviation.

(j) If in Poisson distribution N (P_1) = 932 and m = .02, then N (P_2) shall be 3.92.

[(a) F, (b) F, (c) T, (d) F, (e) T, (f) F, (g) T, (h) F, (i) T, (j) T]

21. (a) What is meant by the theoretical frequency distribution? Discuss the salient features of the Binomial, Poisson and Normal distribution.

    (b) Define Binomial distribution and indicate its chief characteristics. Under what condition does it tend to Poisson distribution?

22. (a) Define Normal distribution. What are the main characteristics of a Normal distribution?

    (b) What is a Binomial distribution? Give a real life example where such a distribution is appropriate.

23. (a) State the properties of normal distribution. It is given that the daily volume of sales in a shop follows a normal distribution with mean of Rs. 5,000 and standard deviation of Rs. 800. Find the percentage of days when the sales will be (i) above Rs. 6,200 and (ii) between Rs. 3,500 and Rs. 6,000.

    (b) Two electric light tubes of manufacturer A have a mean lifetime of 1,400 hours with a standard deviation of 200 hours, while those of manufacturer B have a mean life-time of 1,200 hours with a standard deviation of 100 hours. If random samples of 125 tubes of each brand are tested, what is the probability that the brand A tubes will have a mean lifetime which is at least (i) 160 hours more than the brand B, and (ii) 250 hours more than the brand B tubes?

24. (a) The measurements of the items of a product are approximately normally distributed with a mean of 20 cm. and standard deviation 4 cm. Items which measure between 18 cm. and 23 cm. are sold at 50 paise each and other items at 30 paise each. Find the total amount collected if in all 10,000 items are sold. How many items measure 26 cm. or more, given that

Z	:	.5	.75	1.50
Area	:	.1915	.2734	.4332

(b) Assume that the mean height of a group of men is 64.25 inches with s.d. of 2.81 inches. How many in a group of 400 men would expect to be over 6 ft. tall?

25. A radio shop sells, on an average, 200 radios per day with a standard deviation of 50 radios. After an extensive advertising campaign, the management will compute the average sales for the next 25 days to see whether an improvement has occurred. Assume that the daily sales of radios is normally distributed :

(i) Write down the null and the alternate hypothesis.

(ii) Test the hypothesis at 5% level of significance if $\overline{X} = 216$.

26. (a) Fit Poisson distribution to the following data and calculate the theoretical frequency :

No. of mistakes per page :	0	1	2	3	4
No. of pages on which mistakes occurred :	109	65	22	3	1

(b) In binomial distribution with 6 independent trials, the probabilities of 3 and 4 successes are found to be 0.2457 and 0.0819 respectively. Find the parameter p of the binomial.

27. (a) Why does the normal distribution occupy the most honourable position in statistical analysis?

(b) Describe briefly the characteristics of the normal probability distribution. Why does it occupy such a prominent place in statistics?

28. (a) Explain the role of normal distribution and also point out its constraints.

(b) What are the characteristics of Poisson distribution? Mention three business situations where Poisson model is applicable.

(c) Discuss briefly the importance of normal distribution in statistical theory and its applications.

29. (a) Derive the Poisson probability Law as a limiting form of the binomial probability law.

(b) State the conditions under which the Binomial distribution tends to :

(i) the Poisson distribution, and

(ii) the Normal distribution.

(c) State the conditions which must be fulfilled for using the Binomial Distribution. Also state the parameters of the binomial distribution.

(d) What is Binomial distribution? Under what conditions will it tend to be normal?

20. (a) Of a large group of men, 33% are under 165 cm. in height and 63% are between 160 and 185 cm. Find out the mean and standard deviation of the distribution of height, assuming it to be normal.

(b) In a town 10 accidents took place in a span of 50 days. Assuming that the number of accidents per day follows poison distribution, find the probability that there will be there or more accidents per day.

[0.0012]

21. In 20 trials of an event of small probability, the frequency f of the number 0 successes x is given in the following series :

x :	0	1	2	3	4	5	6
f :	3	2	6	5	5	1	2

The mean number of successes is 2.75. Find the frequencies of the Poisson distribution with the same mean and the same total frequency.

22. (a) Under what conditions can binomial distribution be applied?

A machine produces an average of 20% defective bolts. A batch is accepted if a sample of 5 bolts taken from that batch contains no defective and rejective if the sample contains 3 or more defectives. In other cases, a second sample is taken. What is the probability that the second sample is required?

(b) Assuming that the height distribution of a group of men is normal, find the mean and standard deviation, given that 84% of the men have heights less than 65.2 inches and 68% have heights between 65.2 and 62.8 inches.

33. (a) The I.Q.S. of army volunteers in a given year are normally distributed with mean= 110 and standard deviation=10. The army wants of give advanced training to 20% of those recruits with the highest scores. What is the lowest I.Q. Score acceptable for the advanced training?

(b) The distribution of typing mistakes committed by a typist is given below :

Mistakes per page	:	0	1	2	3	4	5
No. of pages	:	142	156	69	27	5	1

If a Poisson distribution and test goodness of fit.

34. (a) "A binomial distribution need not necessarily be a symmetrical distribution". Do you agree with the statement? Give reasons.
    (b) Define Binomial distribution. State its main properties.
    (c) What is Poisson distribution? Point out its role.
35. (a) Explain the concept of probability distribution. Point out the important properties of normal distribution.
    (b) Write down the probability distribution and its mean and variate. What are the properties of normal distribution?
36. (a) Why does the normal distribution hold the most honourable position in the theory of probability?
    (b) Explain the distinctive features of Binomial, Normal and Poisson probability distributions. When does a Binomial distribution tend to become normal?
37. (a) What are the various constraints of Binomial distribution?
    (b) Show that the mean and variance are identical in a Poisson distribution.
    (c) What is hypergeometric distribution? Explain its properties, Drive the mean and variance of a Binomial distribution.
38. (a) Describe the chief characteristics of normal distribution.
    (b) What are the chief properties of normal distribution? Describe briefly the importance of normal distribution in statistical analysis.
39. (a) Explain the properties of a normal distribution.
    (b) State the conditions for a function of discrete random variable to be a distribution.
    (c) Explain the salient features of binomial and normal probability distribution.
    (d) Prove that the variance of a normal distribution is np q.
40. (a) Derive the mean and variance of a binomial distribution.
    (b) What is normal probability distribution? Explain the characteristic features of a normal distribution.
    (c) What do you understand by hypergeometric distribution? Discuss its properties.

(d) State the conditions under which the binomial distribution tends to:

(i) Poisson distribution, and

(ii) Normal distribution.

41. Which probability distribution is most likely the appropriate one to use for the following variables : Binomial, Poisson or Normal?

(i) The life span of a female born in 1957.

(ii) The number of acts passing through a toll booth.

(iii) The number of defective radio in a lot of 100.

(iv) The water level in a reservoir.

42. (a) Explain the characteristics of Poisson distribution.

(b) Define a Poisson distribution. Mention two situations where the distribution arises.

43. (a) State the important properties of normal distribution. What are its uses in economics?

(b) What is Poisson distribution? Explain with an example and state the conditions under which this distribution is used.

44. (a) State the 'Central limit theorem'. What is its importance?

(b) Define Binomial distribution. What are its chief characteristics?

(c) Explain the distinctive features of Binomial Poisson and Normal Distributions.

45. (a) Derive the mean and variance of a Binomial distribution.

(b) Determine the Binomial distribution for which the mean is 4 and standard deviation is 3.

(c) Name the six situations where Poisson distribution can have applications.

(d) What do you mean by binomial distribution.

46. A typist in a company commits the following number of mistakes per page in typing 432 pages :

Mistakes per page	:	0	1	2	3	4
No. of pages	:	223	142	48	15	4

Fit a Poisson distribution and test the goodness of fit.

47. As a tailor prepared suits for a number of his customers after the preliminary trails and retrials, he had to alter the suits as follows:

No. of trials after preliminary trial	*No. of cases*
0	200
1	75
2	20
3	5
	300

By applying Poisson distribution, calculate the theoretical frequencies.

(b) A factory has two machines. The empirical evidence has established that Machines. I and II produce 30% and 7% of the output respectively. It has also been established that 5% and 1% of the output produced buy these machines respectively was defective. A defective item is drawn at random. What is the probability that the defective item was produced by Machine 1 or Machine 2?

48. (a) (i) The marks of a student in a class are normally distributed with mean 70 and S.D. of 5. If the instructor decides to give 'A' to the top 15% of the class, how many marks a student must get to be able to get 'A' grade?

(ii) If the probability that an individual suffers a bad reaction from an injection of given serum is 0.001, determine the probability that out of 2000 individuals :

(i) exactly three

(ii) more than two individuals will suffer a reaction.

(b) Explain the concept of probability distribution. If three unbiased coins are tossed simultaneously and head noted, prepare a probability distribution and estimate its mean and variance.

49. (a) In a normal distribution 30% of the items are under 50 and 10% are over 86. Find the mean and standard deviation of the distribution area under the normal curve.

(b) The marks of the students are normally distributed. 10% get more than 75 marks and 20% get less than 40 marks. Find the mean and standard deviation of the distribution. The relevant extract of Area table (under the normal curve) is given below:

X	:	0.84	1.28	2.0
Area	:	.2995	.3997	.4772

50. A complex television component has 1,000 joints by machine which is known to produce an average one defective in forty. The components are examined, and faulty soldering corrected by hand. If components requiring more than 35 corrections are discarded, what proportion of the components will be thrown away?

51. (a) Suppose that life of a gas cylinder is normally distributed with a mean of 40 days standard deviation of 5 days. If at a time 10,000 cylinders are issued to customer, how many will need replacement after 35 days.

    (b) In a Binomial distribution consisting of 5 independent trials, probabilities of 1 and 2 successes are 0.4096 and 0.2048. Find the parameter 'p' of the distribution.

52. (a) A large population has a mean height of 150 cm. and a standard deviation of 21 cm. A random sample of size 100 is taken from this population. Find the probability that the sample mean will :

    (i) exceed 151 cm.

    (ii) lie between 148 cm. and 155 cm.

    Assume that the population distribution of height is normal and that sampling is with replacement.

    [(a) 0.4701, (ii) 0.1385].

    (b) The length of telephone calls received by the office of a particular chartered accountant is normally distributed with a mean of 5.3 minutes and a standard deviation of 2.8 minutes. If a single call is selected at random, compute the following probabilities:

    (i) the call will be 2 minutes or less in length ,

    (ii) the call will last between 3 and 5 minutes,

    (iii) the call will be 7 minutes or more in length.

    [(i) 0.1271, (ii) 0.2573, (iii) 0.2611].

53. (a) One per cent of the articles produced by a factory are defective. Assuming that the number of defects follows Poisson distribution, find the probability of getting at least 3 defects in a sample of 100 articles.

    (b) The right-hand grip (in lbs.) of a group of males follows a normal distribution with mean 80.5 lb. and standard deviation 5 lb. Find the probability that a person selected at random will have a grip.

    (i) more than 86.5 lb.

(ii) in the range 40.5 lb. and 82.5 lb.

54. (a) The following table gives the number of days in a 50-day period during which automobile accidents occurred in a city :

Number of accidents	:	0	1	2	3	4
Number of days	:	21	18	7	3	1

Fit a Poisson distribution to the given data.

(b) The time between two arrivals in a queuing model is normally distributed with a mean 2 minutes and standard deviation 0.25 minute. If a random sample of size of 25 is drawn, what is the probability that the sample average will be greater than 2.1 minutes?

(c) Five coins are tossed 96 times :

(i) Construct the theoretical frequency table for 0,1,.....5 heads.

(ii) Find the expected number of times of getting at least 3 heads.

55. (a) Five dice are thrown together 26 times. The number of times 4, 5 or 6 was actually thrown in the experiment is given below. Calculate the expected frequencies.

No. of dice showing 4, 5 or 6	:	0	1	2	3	4	5
Observed frequency	:	1	10	21	35	18	8

[3, 15, 30, 30, 15, 3]

(b) If 30% of the electric bulbs manufactured by a company are defective, find the probability that in a sample of 100 bulbs (a) 0, (b) 1, (c) 2, (d) 3, (e) 4 bulbs will be defective.

[(a) .05, (b) .149, (c) .224, (d) .224, (e) .168]

56. (a) Find the coordinates of the normal curve

(i) z = 2.68, (ii) z = –48, (iii) z = –1.42.

[(i) .0110, (ii) .3555, (iii) .1456]

(b) Find the area under the normal curve between :

(i) z = – 1.5 and z = 2.6

(ii) z = – 1.47 and z = 2.20

(iii) z = – 2.25 and z = 4.0

[(i) 9285, (ii) .0569, (iii) .3324]

57. (a) Find the mean and standard deviation of a normal distribution when 8% of the items are over 64 and 31% are under 45.

(b) Among 10,000 random digits, in how many cases do we expect that the digit 3 appears at most 950 times. (The area under standard normal curve for Z = 1.667 is 0.4525 app.)

58. Past surveys show that 40% of the officers at the certain industry own cars. Suppose six officers are selected at random from this industry with replacement) :

(i) What is the probability that exactly four will own cars?

(ii) What is the probability that at least one will own a car?

(iii) What is the theoretical mean of the probability distribution under consideration?

59. (a) The mean of binomial distribution is 20, and the standard deviation 4. Calculate n, p and q.

[n = 100, p = 1/5, q = 4/5]

(b) Out of 100 families with 5 children each how many would you expect to have (a) 3 boys, (b) 5 girls, and (c) either 2 or 3 boys? Assume equal probabilities for boys and girls.

[(a) 250, (b) 25, (c) 500]

60. (a) A manufacturer finds that one article in every twenty is below the required standard. How many sub-standard articles would we expect to find in a sample of 200?

[7 to 13]

(b) An electric bulb manufacturer finds that 4 per cent of the bulbs are defective. How many perfect bulbs are to be expected in a sample of 400? What are the chances that a random sample of 10 will not contain more than one defective bulb?

[380 to 388; 0.89]

61. Proof reading of 200 pages of a book containing 500 pages gave the following results :

No. of mistakes per page :	0	1	2	3	4	5
Frequency :	113	62	20	3	1	1
Cost per page of checking :	1.0	1.5	2.5	3.0	3.5	4.00

(a) Fit a Poisson distribution and estimate the total cost of correcting the whole book.

X	:	0	1	2	3	4	5
f	:	110	66	20	4	1	0

62. (a) The screws produced by a certain machine were checked by examining number of defectives in a sample of 12. The following table shows the distribution of 128 samples according to the number of defectives items they contained :

No. of defectives in sample of 12	:	0	1	2	3	4	5	6	7	Total
No. of samples	:	7	6	19	35	30	23	7	1	128

(a) Fit a Binomial distribution and find the expected frequencies if the chance of machine being defective is 1/2.

(b) Find the mean and standard deviation of the fitted distribution.

(b) The average daily sales of 500 branch offices was Rs. 1,50,000 and the standard deviation Rs. 15,000. Assuming the distribution to the normal, indicate how many branches have sales between

(i) Rs. 1,20,000 and Rs. 1,45, 000

(ii) 1,40,000 and Rs. 1,65,000.

63. A normal distribution has mean, $\mu = 12$ and $\sigma = 2$. Find the following area under the curve:

(a) From x = 10 to x = 13.5

(b) From x = 11.4 to x = 14.2

(c) From x = 9.6 to x = 13.8

(d) From x = 6 to x = 18.

[(a) .617, (b) .4822, (c) .7008, (d) .9973]

64. If the height of 300 students is normally distributed with mean 68 inches and standard deviation 2 inches, how many students have height :

(a) greater than 72 inches,

(b) less than or equal to 60 inches,

(c) between 65 and 71 inches inclusive, and

(d) equal to 68 inches?

Assume the measurements to be recorded to the nearest inch.

[(a) 20, (b) 28, (c) 205, (d) 40]

65. If the height of 12,000 college men closely follow a normal distribution with mean 69 inches and the standard deviation 2.5 inches, answer the following:

    (a) How many of these men would you expect to be at least 6 feet in height?

    (b) What range of height would you expect to include the middle 75% of the mean in this group?

    [(a) 1,151, (b) 66.125 inches to 71.875 inches]

66. (a) A certain automatic screw manufacturing machine produces on an average one slotless screw among every 100 screws. If the screws are packed in boxes of 300, what percentage of these boxes would you expect to have (i) no slotless screw, and (ii) at least one slotless screw?

    [(i) 4.98%, (ii) 95.02%]

    (b) In an examination 15% of the candidates got first class (60% marks or above) While 40% failed (securing below 40%). Assuming the marks to be normally distributed. Estimate the mean and standard deviation.

    [$\mu = 43.94, \sigma = 15.5$]

67. (a) A survey of male children in 128 families each having 5 children gave the following data:

No. of male children :	0	1	2	3	4	5	Total
No. of families :	9	17	26	39	22	12	125

Fit a Binomial distribution of the data :

[(a) 2.74, 15.70, 36.03, 41.36, 23.72, 5.45, (b) $\mu_2 = 2.275$, $\mu_3 = 0.6825$, $\mu_4 = 147$].

68. (a) What are the properties of normal curve?

    (b) Prove that the mean of binomial distribution is np.

    (c) Describe the chief characteristics of Poisson distribution.

    (d) Find the mean and variance of Poisson distribution.

    (e) What are chief characteristics of normal distribution? Under what circumstances does the binomial tend to the normal distribution?

69. Records show that the probability is 0.00002 that a car will have a flat tyre while driving over a certain bridge. Use the Poisson probability

distribution to determine the probability that among 20,000 cars driven over this bridge, nor more than one will have a flat tyre.

70. Etona Co. Ltd. manufacturing staple pins, markets its product in packets of 1,000 and there is a small chance, 0.001 of a staple pin to be defective. The company guarantees not more than four defective pins in each packet.

    (i) What is the probability that a packet will meet the guarantee?

    (ii) If the Company sells 10,000 packets per month, what is the expected number of guarantee claims in a month?

71. (a) What probability model is appropriate to describe a situation where 100 misprints are distributed randomly throughout the 100 pages of a book? For this model, what is the probability that a page observed at random wil contain at least three misprint?

    (b) One hundred car stereos are inspected as they came off the production line and number of defects per set is recorded below:

No. of defects	:	0	1	2	3	4
No. of sets	:	79	18	2	1	0

    Fit a Poisson distribution to the above data.

72. (a) Discuss the salient features of Normal distribution. Explain the characteristics of Poisson distribution.

    (b) Explain how for the normal distribution, the mean, median and mode are equal.

73. (a) How is the probability distribution of a discrete random variable defined? Define its expected value.

    (b) List the chief properties of the normal distribution. Why is this distribution given a central place in statistics?

    (c) Describe normal distribution and discuss its properties. Why is it so important in behavioural sciences?

74. (a) Define Poisson distribution and state the conditions under which it is used.

    (b) Explain briefly the characteristics of the Binomial and Poisson distributions. How are their means and variances calculated?

75. (a) What are the assumptions of Poisson distribution? Explain the distribution, stating its mean and variance.

    (b) What is Poisson Distribution. Distinguish clearly the relationship between Binomial Poisson Distribution.

76. In a certain book, the frequency distribution for the number of words per page may be taken as approximate normal with mean 800 and standard deviation 50. If 3 pages are chosen at random, what is the probability that none of them has between 830 and 845 words each?

[0.825]

77. What is Normal distribution? Describe its properties in detail. Bring out its importance in statistics.

78. (a) The mean and standard deviation for the lifetimes of a population of light bulbs are 1,200 and 150 hours, respectively. Assuming these lifetimes are normally distributed, what is the probability that a light bulb will last over 1,500 hours?

(b) In a certain examination of Delhi University, the mean mark was 48, and 12.7% of the candidates failed (*i.e.*, below 40). Assuming a normal distribution of marks, find the percentage of candidates who secured first division.

79. (a) Suppose that life of a gas cylinder is normally distributed with a mean of 40 days and a standard deviation of 5 days. If at a time, 10,000 cylinders are issued to customers, how many will need replacement after 35 days?

80. (a) Suppose that the life of a gas cylinder is normally distributed with a mean of 40 days and a standard deviation of 5 days. If, at a time, 10,000 cylinders are issued to customers, how many will need replacement after 35 days.

(b) A machine fills milk into milk cartons such tat the mean of the distribution of fills is 1 litre with a standard deviation of 10 ml. If the distribution of fills is normal :

(i) What percentage of the filled up cartons will contains less than 1 litre?

(ii) What percentage of the filled up cartons will contains less than 975 ml.? More than 1010 ml.?

(d) Prove that the binomial distribution is symmetrical when p=1/2.

81. Tick the correct answer :

(a) The standard deviation of Binomial distribution is :

(i) $\sqrt{nqp}$, (ii) npq, (iii) $n^2p^2q^2$, (iv) npq/6pq, (v) np.

(b) In case of Binomial distribution β_1 is

(i) (q-p)/npq_2, (ii) (q-p)/np, (iii) (q-p)/npq, (iv) npq/(q-p), (v) npq/$(p-q)_3$.

(c) The fourth moment of Poisson distribution is

(i) m+3m, (ii) 3m, (iii) m_2+3m, (iv)m+m_2, (v)m_2+3.

(d) The standard deviation of Poisson distribution is :

(i) m, (ii) 3m, (iii) 4.2, (iv) 6.8, (v) none of these.

(e) If in case of Poisson distribution $\mu 2 = 3.2$, $\mu 3$ will be

(i) 2.4, (ii) 3.4, (iii) 4.2, (iv) 6.8, (v) none of these.

(f) In case of normal distribution $\beta 2$ is

(i) zero, (ii) 2, (iii) less than 3, (iv) 3, (v) greater than 3.

(g) In case of normal distribution $\overline{X} \pm 2\sigma$ covers

(i) 25.45%, (ii) 95.54%, (iii) 94.45%, (iv) 99.70%, (v) 99% items.

(h) If in Binomial distribution mean is 10 and standard deviation 2, q will be

(i) 0, (ii) 1, (iii) 04, (iv) 0.8, (v) none of these.

(i) If mean of Poisson distribution is 8, μ shall be

(i) 8, (ii) 8^2, (iii) 200, (iv) 3,8, (v) none of these.

(j) In case of normal distribution $\mu 4$ =

(i) 3, (ii) $3\sigma_4$, (iii) $3\sigma^2$, (iv) 0, (v) none of these.

[(a) (i), (b) (iii), (c) (iv), (d) (ii), (e) (iii), (f) (iv), (g) (i), (h) (iv), (i) (iii), (j) (ii)]

82. Fill in blanks :

(a) The probability distribution is the outcome of different ....taken by the random variable X.

(b) Binomial distribution is associated with the name of ....mathematician........

(c) ....was originated by the French Mathematician Simeon Denis Poisson in ....

(d) In a Poisson distribution m4 = ...and b2 = ...

(e) The normal distribution is an approximation to ....

(f) In a normal distribution b1 = ...and b2 = ...

(g) In a normal distribution the points of inflexion occur at ....

(h) The.......... deviation is 4/5 of ....

(i) The normal distribution with X = 0 an f = 1 is known as ....

(j) The mean plus and minus 1.96 standard deviation includes........ percent of the items in normal distribution.

[(a) Probabilities, (b) French, James Benoulli, (c) Poisson distribution, 1837, (d) m $\pm$ 3m^2. 3 + 1/m, (e) Binomial distribution, (f) 0.3, (g) 0.8X $\pm$ f, (h) mean, standard deviation, (i) standard normal distribution, (j) 68.27]